ENVIRONMENT AND MAN
VOLUME SIX

The Chemical
Environment

ENVIRONMENT AND MAN: VOLUME SIX

Titles in this Series

Volume 1   *Energy Resources and the Environment*

Volume 2   *Food, Agriculture and the Environment*

Volume 3   *Health and the Environment*

Volume 4   *Reclamation*

Volume 5   *The Marine Environment*

ENVIRONMENT AND MAN
VOLUME SIX

The Chemical Environment

General Editors

John Lenihan
O.B.E., M.Sc., Ph.D., C.Eng., F.I.E.E., F.Inst.P., F.R.S.E.

Director of the Department of Clinical Physics and Bio-Engineering, West of Scotland Health Boards; Professor of Clinical Physics, University of Glasgow; Chairman of the Scottish Technical Education Council.

and

William W Fletcher
B.Sc., Ph.D., F.L.S., F.I.Biol., F.R.S.E.

Professor of Biology and Past Dean of the School of Biological Sciences. University of Strathclyde; Chairman of the Scottish Branch of the Institute of Biology; President of the Botanical Society of Edinburgh.

1977

ACADEMIC PRESS·NEW YORK & SAN FRANCISCO
A Subsidiary of Harcourt Brace Jovanovich, Publishers

Blackie & Son Limited
Bishopbriggs
Glasgow G64 2NZ

450/452 Edgware Road
London W2 1EG

© 1977 Blackie & Son Ltd·
First published 1977

International Standard Book Number
0-12-443506-8
Library of Congress Catalog Card Number
75-37435

Printed in Great Britain by
R. MacLehose & Co. Ltd.

Background to Authors

Environment and Man: VOLUME SIX

HUMPHRY J. M. BOWEN, M.A., D.Phil., is Reader in Analytical Chemistry at the University of Reading. He is also interested in Botany and Lichenology, and is involved in practical conservation of the environment through his local Naturalists' Trust.

LEONARD J. GOLDWATER, A.B., M.S.P.H., M.D., Med.Sc.D. is Chief of Occupational Medicine at Duke University, Durham, North Carolina. From 1948 to 1969 he was Professor of Occupational Medicine at Columbia University, retiring with Emeritus status in 1969. He has served as a consultant to government agencies, trade unions and industries—and in particular to the World Health Organization and the International Labour Office. He received the Robert A. Kehoe Award of Merit from the American Academy of Occupational Medicine in 1975.

WOODHALL STOPFORD, B.A., B.M.S., M.D., is Asssistant Professor at the School of Public Health, University of North Carolina. He is a Consultant in Occupational Medicine and Toxicology.

MICHAEL R. MOORE, B.Sc., Ph.D., is Lecturer in Materia Medica at the University of Glasgow. For the past nine years he has studied lead toxicology and the porphyrias, and has written extensively on these topics.

BRIAN C. CAMPBELL, M.B., Ch.B., M.R.C.P.(U.K.), is Lecturer in Materia Medica at the University of Glasgow. At present he is engaged on a major survey of environmental lead pollution in Scotland.

ABRAHAM GOLDBERG, M.D., D.Sc., F.R.C.P., F.R.S.E., is Regius Professor of Materia Medica at the University of Glasgow. He is Chairman of Grants Committee 'A' of the Medical Research Council. From 1967 to 1972 he was Director of the MRC Iron and Porphyrin Group at the Western Infirmary, Glasgow, and latterly at Stobhill General Hospital, Glasgow.

M. D. KIPLING, V.R.D., M.A., M.R.C.S., D.P.H., is Regional Employment Medical Adviser in the Employment Medical Advisory Service, U.K., and Honorary Research Fellow in the Department of Social Medicine at the University of Birmingham. From 1950 to 1955 he was Works Medical Officer in Imperial Chemical Industries Ltd., Billingham, and in 1974 he was Joseph Henry Lecturer at the Royal College of Surgeons.

C. ALLEN LINSELL, M.B., B.S., D.C.P., F.R.C.Path., is Chief of the Interdisciplinary Programme and International Liaison at the International Agency for Research on Cancer, Lyon, France. He was previously Director of Laboratory Services in Nairobi, Kenya.

MURIEL L. NEWHOUSE, M.D., F.R.C.P., is Reader in Occupational Medicine at the Trades Union Congress Centenary Institute of Occupational Health, London School of Hygiene and Tropical Medicine. She has served on numerous committees concerned with the asbestos hazard, and has written extensively on the subject.

Series Foreword

MAN IS A DISCOVERING ANIMAL—SCIENCE IN THE SEVENTEENTH century, scenery in the nineteenth and now the environment. In the heyday of Victorian technology—indeed until quite recently—the environment was seen as a boundless cornucopia, to be enjoyed, plundered and re-arranged for profit.

Today many thoughtful people see the environment as a limited resource with conservation as the influence restraining consumption. Some go further, foretelling large-scale starvation and pollution unless we turn back the clock and adopt a simpler way of life.

Extreme views—whether exuberant or gloomy—are more easily propagated, but the middle way, based on reason rather than emotion, is a better guide for future action. This series of books presents an authoritative explanation and discussion of a wide range of problems related to the environment, at a level suitable for practioners and students in science, engineering, medicine, administration and planning. For the increasing numbers of teachers and students involved in degree and diploma courses in environmental science the series should be particularly useful, and for members of the general public willing to make a modest intellectual effort, it will be found to present a thoroughly readable account of the problems underlying the interactions between man and his environment.

Preface

IN THIS VOLUME SIX AUTHORS DISCUSS HAZARDS ARISING FROM THE chemical environment—reviewing historical developments, harmful effects, control measures and future prospects. Studies of this kind are important, because man is unique among living creatures in the ability to control and modify his environment, and timely, because this power is rapidly being augmented by advances in technology, and is not always used with proper regard for the consequences.

The subject also reminds us forcefully that we cannot stand aside from our environment. Each of us is a temporary perturbation resulting from the rearrangement of atoms which were there before we came and will be there after we have gone. A century ago this last sentence could have been written in terms of molecules rather than atoms—but today we are, at an increasing rate, changing the chemical environment by adding materials which have no place in the natural order of things.

Our ingenuity in modifying the chemical environment is accompanied by growing anxiety over the harmful consequences of such activity, and by sharper awareness of the toxic hazards that surround us in everyday life. Nature is by no means pure and wholesome. Dr H. J. M. Bowen reminds us that great quantities of poisonous materials are continually being recycled by geological, meteorological and biological processes. Technology has given us the power to interfere with these equilibria, even though we do not adequately understand them. Dr. Bowen gives a comprehensive account of the natural cycles which regulate the distribution of many important materials, and shows how they are being disturbed—often with detrimental effects—by the activities of man.

Dr. L. J. Goldwater reviews the long history of mercury as a pollutant of the external and internal environments. In well-reasoned arguments, he offers a sceptical view of the much-publicized concern over the presence of methyl mercury in fish. His analysis of this issue shows that chemical hazards cannot be adequately assessed by measuring concentrations in the biosphere; scientific study, often needing much biological insight, is the only basis for rational judgment.

Lead is another metal with a long history. Its toxic hazard, recognized since ancient times, seemed until recently to be well controlled; but, as Dr M. R. Moore and his colleagues explain, the dramatic drop in notifications of industrial lead poisoning since the beginning of the present century has been accompanied by a steep rise in the input of lead to the environment, mainly from motor fuel. More recently, the possible effects on the general public of continual exposure to quite small amounts of lead have generated justifiable anxiety. Dr. Moore and his colleagues give a detailed account of the industrial uses and biological effects of lead, and suggest how the hazards should be controlled.

Dr M. D. Kipling surveys the colourful history of arsenic in industry and medicine. Though no longer favoured as a therapeutic agent or an instrument of homicide, arsenic is so widely useful in agriculture (as a Pesticide and a defoliant), electronics (in the manufacture of semiconductors) and in many other industries that its dangers continue to provide difficult problems. Its role as a carcinogen is still debated, and there is evidence that it may actually be beneficial in some animals.

Dr C. A. Linsell relates an intriguing story of detection which began in 1960 when 100,000 turkeys died mysteriously in England. Soon there were reports from many other countries of heavy mortality among chickens, partridges and trout. Massive scientific resources were mobilized to attack the problem. By 1962 it was known that the animals had died of liver cancer, induced by aflatoxins. These compounds, produced by fungi closely related to *Penicillium*, may occur in peanuts and, to a lesser extent, in many other nuts, seeds and cereals. In the developed countries of the world, monitoring techniques provide adequate control of the hazard—but in the countries of the Third World, where the processing and inspection of food are less sophisticated, the dangers are greater.

Anxiety arises also because it seems that aflatoxins may be converted to more dangerous substances by the metabolic processes of man and other animals. The story is not yet finished, but already it provides an interesting example of the difficulties involved in identifying and controlling a long-standing environmental hazard.

Dr. M. L. Newhouse explains how asbestos constitutes a major health hazard in many industries and a possible risk to city dwellers. Asbestos, the only naturally occurring fibrous material, is so widely used—especially in the construction industries—that it would be difficult to replace. But inhalation of the fibres or dust can lead to cancer (after a latent period of as long as 20 years) or to chronic lung disease. Dr Newhouse describes the epidemiological methods used in the patient unravelling of the links between asbestos and disease, and discusses the preventive measures needed to keep the risks within acceptable limits in the future.

The six chapters all convey the excitement of research into complicated and often urgent problems. In different ways, they all underline the message that conscientious understanding based on scientific study is the essential preliminary to adequate management and control of the chemical environment.

Contents

CHAPTER ONE— NATURAL CYCLES OF THE ELEMENTS   1
AND THEIR PERTURBATION BY MAN
by H. J. M. Bowen

Residence times in the atmosphere. Residence times in water.
Residence times in soils. Cycling of elements in the environment.
Perturbations of elementary cycles. The carbon cycle. The oxygen
cycle. The nitrogen cycle. Reduction of dinitrogen to ammonia.
Oxidation of ammonia to nitrate. Conversion of nitrates to dinitrogen.
Production of oxides of nitrogen. The sulphur cycle. The phosphorus
cycle. the fluorine cycle. Heavy metal cycles. Copper, lead, zinc,
arsenic, cadmium, chromium, mercury, nickel. Summary. History of
perturbations of natural cycles. Further reading.

CHAPTER TWO— MERCURY   38
by Leonard J. Goldwater and Woodhall Stopford

Introduction. General and physical properties. Occurrence. Sources.
Forms and compounds. Chemistry. Mercury release (natural and
man-made). Analytical methods. Toxicological and physiological
aspects (including risks to man). Specific toxicological effects. Metallic
mercury. Metallic vapour. Inorganic salts. Organic compounds.
Arylmercurials. Alkoxyalkylmercurials. Alkylmercurials. Methyl
mercury in fish. The status of methyl mercury. Analytical difficulties.
Methylation of mercury at Minamata. Human poisoning by methyl
mercury. Methylation and demethylation of mercury. Nature's
protective mechanisms. Development of tolerance. Genetic damage.
Human intake of mercury from food and other sources.
Environmental standards for mercury. Bases for standards.
"Normal" values for mercury. Water. Drinking water. Waste water.
Air, Industrial air. Ambient air. Food. Conclusions. Further reading.

CHAPTER THREE— LEAD   64
by M. R. Moore, B. C. Campbell and A. Goldberg

Introduction. History. Human exposure. Sources of lead exposure,
Food. Water. Air. Lead in soils and plants. Lead poisoning in animals.
Lead poisoning in man. Clinical lead poisoning. Acute lead
poisoning. Chronic lead poisoning. Industrial lead exposure.
"Moonshine" whisky. Domestic water. Children. A common picture
of lead poisoning in the adult. Organic lead poisoning. Presentation in
children. Pathological effects of lead. The blood. The nervous system.
The kidney. Biochemical effects of lead. Sub-clinical or biochemical
lead poisoning. Prevention and treatment—prophylaxis. Treatment of
lead poisoning. Further reading.

CHAPTER FOUR— ARSENIC   93
by M. D. Kipling

Arsenic and its compounds. Tests for arsenic. Arsenic in the environment. Arsenic in food. Arsenic in man. The uses of arsenic. Historical. Contacts with arsenic. Arsenic poisoning. Effects of arsenic. Causes of poisoning. Industrial poisoning. Water pollution. Arsenic poisoning—homicidal. Arsine poisoning in industry. Ships and poisoning. Arsenic and cancer. Skin cancer. Internal cancer. Animals. Protection. Summary. Further reading.

CHAPTER FIVE— AFLATOXINS   121
by C. A. Linsell

The toxic cause of turkey X disease. The occurrence of the aflatoxins in nature. The chemical composition and properties of the aflatoxins. The metabolism of aflatoxins. The acute toxicity of aflatoxin. Acute toxicity in man. The carcinogenicity of the aflatoxins. The geographic distribution of human liver cancer. Liver cancer in Africa and Asia. Liver cancer in Europe and North America. Cirrhosis and liver cancer. Liver cancer in the Tropics and aflatoxin. Conclusion. Further reading.

CHAPTER SIX— ASBESTOS   137
by M. L. Newhouse

Structure and uses. Biological effects. Asbestos corns. Asbestos bodies. Asbestosis. Asbestos plaques. Asbestos-related cancers. Cancer of the lung. Cancers of the pleura and peritoneum—the mesotheliomata. The size of the risk. Other asbestos-related tumours. Mechanism of asbestos cancer production. The risk to the community. Environmental pollution Asbestos in food, water and other beverages. The finished product. Assessment of the environmental risk. Future policy and research. Further reading.

INDEX   159

CHAPTER ONE

NATURAL CYCLES OF THE ELEMENTS AND THEIR PERTURBATION BY MAN

H. J. M. BOWEN

ONE OF THE GENERALIZATIONS ARISING FROM STUDIES OF environmental science over the last twenty years is that static equilibria are very rare on the earth's crust. For many years the atmosphere and the oceans were regarded as static systems. We now recognize that both these systems are actually in dynamic equilibrium, and that of their components probably only neon, krypton and xenon are invariant with respect to time. Another part of the external environment, the soil, is also a dynamic rather than a static system; the same may be said of underwater sediments. The continual fluxes which occur in all parts of the earth's crust contrast markedly with the static conditions which appear to obtain on the surface of the moon.

A dynamic situation is characterized by inputs, outputs and residence times for its constituents. At equilibrium, input and output rates are equal, and concentrations within the system remain constant. It is not possible to distinguish static equilibria from dynamic equilibria by measurements of concentrations alone. Actual measurements of input or output rates are needed to prove that the system is dynamic. Alternatively, the introduction of radioactive nuclides into the system is a powerful method of measuring input and output rates without disturbing the equilibrium.

A simple example of a system in dynamic equilibrium is a domestic bath with the tap left running and the plug removed. Provided that the rate of input of water is equal to its rate of loss through the plughole, the water level in the bath will remain constant. For such a system, the *residence time* of any substance X can be defined in the following way:

1

$$\text{Residence time} = \frac{\text{mass of X in the system}}{\text{rate of input of X into system}}$$

where the denominator may be either the rate of input or output, since these are identical at equilibrium. X may be water or anything dissolved in the water. Different constituents may have different residence times in the same system. For example, if the bath was being replenished with muddy water, the mud could settle out in the bath, and the residence time of the mud would be longer than that of the water. The bath is then said to be a trap for mud. implying an increased residence time.

Rather surprisingly, the idea of residence times in natural systems is of comparatively recent origin. The term was first used by Barth in 1951 but, despite its wide applicability, not many residence times have been measured. The residence time is the reciprocal of the *rate constant* for input or output, which may be a more familiar term to chemists. It is 1.44 times the *half-life*, which is the time needed for half the molecules present to be lost from the system. If a molecule is lost by two different processes at different rates, the overall residence time t is given by:

$$\frac{1}{T} = \frac{1}{T_1} + \frac{1}{T_2}$$

where T_1 and T_2 are residence times for the two outputs considered individually.

Residence times in the atmosphere

Apart from the inert gases, which may be static or may be accumulating as products of radioactive decay, there are three ranges of lifetime for gases in the atmosphere. The two major components, nitrogen and oxygen, are long-lived, with residence times of about a million and ten thousand years respectively. They have biologically-mediated inputs and outputs, and may only be in pseudo-equilibrium, as they could still be accumulating on a geological time scale. Medium-lived and short-lived gases in the atmosphere are listed in Table 1.1. Since the mixing time of the troposphere (the atmosphere below an altitude of 12 km) is relatively slow, residence times of less than a year are determined by local mixing conditions, and may vary from place to place. The residence times given in Table 1.1 are recent estimates, but are not known with any great precision.

A characteristic feature is that variability of composition is inversely proportional to residence time. Thus all the short-lived gases in Table 1.1 have very variable concentrations in the atmosphere. Medium-lived gases are more constant, but daily variations in carbon dioxide can just be measured. No-one has been able to detect any variation in the concentration of either oxygen or nitrogen.

Table 1.1 Inputs, outputs and residence times of some gases in the atmosphere

Gas	Input	Output	Residence time	
CO_2	animal/microbial	plant	5 years	
CO	microbial	microbial?	2 years	
CH_4	microbial	microbial	5 years	medium-lived
H_2	microbial	microbial	5 years	
N_2O	microbial	microbial	10 years	
H_2O	evaporation	rainfall	10 days	
NH_3	microbial	microbial/rain	2 days	
SO_2	volcanoes+	rainfall+	4 days	short-lived
NO_2	microbial+	rainfall+	3 days	

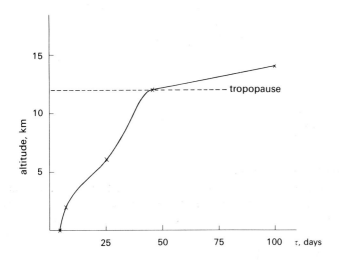

Figure 1.1 Variation of residence time with altitude for particles in the atmosphere.

Residence times of particulate matter in the atmosphere vary with height (figure 1.1). It is generally assumed that particles which are too small to settle out under gravity are efficiently scavenged by rainfall. Assuming a mean rainfall of 1 mm per day, calculations show that the residence time of a particle in the troposphere should be 24 days, which agrees rather well with actual measurements (22–40 days).

Residence times in water

While methods for determining residence times in freshwater systems are known, they have been applied in very few cases. The lack of data means that generalizations are of little use, so we will outline three typical cases: a river, a lake and an estuary.

For the river, consider the non-tidal part of the River Thames above Teddington lock. This river has a length of 237 km, a mean cross-sectional area at Teddington of 186m^2, and its mean output rate is 1.04×10^6 m^3/hour. We can estimate the instantaneous volume of water in the Thames by treating it as a cone bisected down its long axis; this volume comes to 14.7×10^6 m^3, so the residence time for water is given by

$$\frac{\text{volume}}{\text{flow rate}} = \frac{14.7 \times 10^6}{1.04 \times 10^6} = 14 \text{ hours}$$

This relatively short residence time for water shows how it is that rivers can purify themselves so rapidly. It also tells us that many pollutants may not have time to reach equilibrium, with respect to their distribution between water and mud particles, before they reach the sea.

Lake Washington, near Seattle in the United States, has been carefully studied for some years. Here the mass of water in the lake is 2.9×10^{12}kg, the mean outflow rate is 3.2×10^9 kg/day, so that the residence time of water is

$$\frac{\text{mass}}{\text{flow rate}} = \frac{2.9 \times 10^{12}}{3.2 \times 10^9} = 906 \text{ days}$$

Estimates have also been made of the residence times of various elements in this lake, as follows: sodium, 906 days; arsenic, copper, mercury and zinc, 340–560 days; iron and lead, 12–25 days. Sodium is entirely removed in the outflowing water, but iron and lead rapidly precipitate onto the lake sediments; the other elements are intermediate in behaviour. Hence Lake Washington acts as a sink trap for any iron and lead added to the streams feeding it. It is not a simple matter to obtain residence times for lakes, since the following inputs and outputs should be measured:

Inputs	*Outputs*
1. Streams flowing in	5. Streams flowing out
2. Groundwater flowing in	6. Groundwater flowing out
3. Rainfall	7. Evaporation
4. Back diffusion from sediments	8. Loss to sediments

Items 1, 3, 5, 7 and 8 can usually be measured, but need regular monitoring over at least one year to obtain average values; items 2, 4 and 6 are exceedingly difficult to measure and are usually neglected.

For water in estuaries, the calculation of residence times is quite complicated, because of the ebb and flow of the tide, the different densities of fresh and salt water, and the funnel-shaped geometry. However, it appears that for estuaries without marked constrictions, the residence time of water is measured in days; a recent calculation for the Mersey estuary gave 5.3 days. The chemical changes in the dissolved constituents and particles which occur when fresh water meets the sea are not well understood. However, it is clear that many constituents of river water finish up in the estuarine muds and are not discharged to the ocean.

The ocean itself can be considered as a very large lake with inputs from the world's rivers and outputs to marine sediments. It is fairly easy to obtain the following formula for the residence time T_X of any constituent X dissolved in sea water:

$$T_X = \frac{38380\,[X]_{\text{sea}}}{[X]_{\text{river}}} \quad \text{years}$$

where the [X]s are concentrations in identical units, and 38380 years represents the residence time of water in the ocean. The concentrations of most of the chemical elements in sea water are known, but the mean concentration in the global river input is much less easily obtained.

An alternative expression for residence times in the ocean can be obtained from output data. A constant rate of output by sedimentation is assumed, and the ultimate product is sedimentary rock, whose composition is fairly well known. It can be shown that

$$T_X = \frac{4.42 \times 10^8 [X]_{\text{sea}}}{[X]_{\text{rock}}} \quad \text{years}$$

Here 4.42×10^8 years is a representative residence time of marine sediments. The two expressions give comparable values for residence times of most common elements, but the discrepancies become marked for some rare elements whose chemical analysis presents problems. Current estimates of some residence times in the ocean are given in Table 1.2.

Table 1.2 Approximate residence times in the ocean

Residence time range	Elements
0.1— 1 kiloyears	Al, Fe, La, Pb, Th
1— 10 kiloyears	Cu, Cr, Mn, Sn
10— 100 kiloyears	Ag, Ba, Co, Hg, Ni, Sb, Se, Si, V, Zn
0.1— 1 megayears	As, Au, Cs, F, I, Mo, P, W
1— 10 megayears	Ca, Li, K, Rb, Sr, U
more than 10 megayears	B, Br, Cl, Mg, Na

1 kiloyear = 1000 years. 1 megayear = 1 000 000 years

Thus elements such as aluminium, iron and silicon, which are major components of marine sediments, and also elements which concentrate in manganese nodules, have residence times less than the residence time of water in the ocean. The concentrations of many of these elements often vary with location or depth. On the other hand the alkali and alkaline-earth metals, the halogens and boron have very long residence times in the ocean, and have presumably accumulated there over geological time. No variations in composition, other than those caused by varying salinity, have been reported for these long-lived elements in the sea.

Residence times in soils
Soil inputs include rainfall, dry dustfall, fertilizers and weathering of parent rock. Soil outputs are mainly by drainage or cropping, or in some cases by wind erosion. Very few studies have been made on specific soils, but those that have suggest that the residence times of most elements in soils are of the order of hundreds or more often thousands of years. Some of the labile elements in soils, such as sodium and chlorine, may have residence times as low as a few tens of years, but the majority of elements, including potassium, phosphorus and most heavy metals bind strongly to one or more component of the soil and have long residence times. The organic humus in the soil can be dated from its content of radioactive carbon–14, and different organic fractions can be isolated with ages in the range 50–1400 years.

Cycling of elements in the environment
There are two kinds of cycles involving chemical elements which occur in the environment. The first kind is geological, and the most obvious example is the cycle of water, some aspects of which have been mentioned above. Radiant energy from the sun evaporates water from the ocean, some of which falls as rain on land. The rainwater both dissolves rocks and physically transports rock fragments and mud as it returns to the sea. Meanwhile elements are slowly precipitated from the ocean to form sediments of various kinds. Buried sediments become converted to hard rocks by pressure, and these rocks may later be uplifted by earth movements to form dry land. A more localized geological cycle involves volcanic emissions. Here molten rock either flows out to cover the surrounding terrain, or is ejected with explosive force into the atmosphere. Volcanic dust which reaches the stratosphere is distributed by winds and may fall back on far-distant parts of the earth. All the elements are involved in geological cycles, but the most important by weight are oxygen, silicon, aluminium, iron, calcium, magnesium, sodium and potassium.

The second kind of cycle is biological. Living organisms are able to use radiant energy from the sun to carry the chemical elements through a range of cyclical processes. The main elements involved are carbon, oxygen, hydrogen, nitrogen and sulphur. A striking feature of biological cycles is their capacity for oxidation and reduction. Thus the basic biological reaction of photosynthesis involves the oxidation of water to molecular oxygen. Most other outputs of biological cycles turn out to be reduced forms of the primary inputs. Thus carbon dioxide, a primary plant food and therefore an input, ultimately appears in part as two reduced products or outputs of long residence time: coal and oil. In the same way some of the nitrates used as inputs are reduced to molecular nitrogen, and some sulphate appears as elementary sulphur. These products are visible outputs of the biological cycles involved.

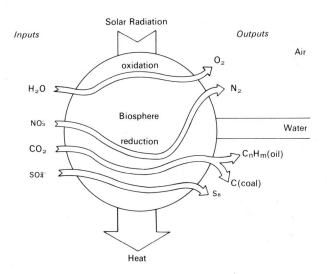

Figure 1.2 Diagram of major inputs to, and outputs from, the biosphere.

The overall processes occurring in the biosphere are shown in simplified form in figure 1.2. This figure makes no effort to portray the exceedingly complicated cycling and recycling of elements within the biosphere. Most scientists agree that all coal, oil, atmospheric oxygen and "salt-dome" sulphur deposits are outputs of the biosphere, but it is not clear what proportions of atmospheric nitrogen have been produced by biological as distinct from geological processes.

Perturbations of elementary cycles
Planetary domination by man has led to a number of perturbations of natural cycles. In the first place, man has increased the input rates of several

substances, for example, carbon dioxide, sulphur dioxide and lead to the atmosphere. Assuming that residence times are constant, a consequence of increased inputs must be increased concentrations of these substances in environmental reservoirs. Thus carbon dioxide is increasing in concentration in the atmosphere, and sulphur and lead are increasing in the upper layers of soils. These increases in concentration are often the first or only evidence that man is significantly altering natural input rates, since the rates themselves are more difficult to measure.

Secondly, man is removing substantial quantities of biological outputs, such as coal, oil and sulphur, at rates much greater than their natural rates of replenishment. As will be seen below, the rates of replenishment are seldom precisely known, and very little effort has been devoted to estimating them. The policy of cashing in on resources which have taken millions of years to form, is one which future generations will rightly deplore. In an ideal world, rates of production and consumption of resources should balance.

Thirdly, man is modifying the biosphere over a large percentage of the surface area of available land, by deforestation, agriculture and intensive grazing. It is difficult to estimate the effects of these changes in the biosphere, but it seems likely that both the total content of elements in the biosphere and their residence times may be affected. Residence times in the

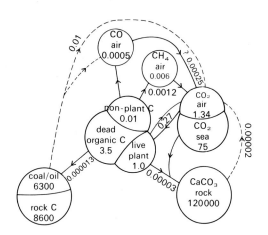

Figure 1.3 Outline of the carbon cycle. Circles represent reservoirs of carbon, and the figures within them represent kg/m² earth's surface. Lines with arrows represent natural processes, and the figures attached to the arrows are estimated fluxes in kg C/m^2 yr. Dotted lines represent human perturbations of the cycle. The total surface area of the earth is $5.12 \times 10^{14}m^2$.

iving biosphere are usually of the order of a few years or tens of years, which are longer than the single season involved with a cereal crop. Modification of the terrestrial part of the biosphere can be expected to lead to short-term changes in the rates of cycling of elements and much longer-term changes in soils. Human perturbation of the marine biosphere, where residence times may be as short as a few hours, has been far less obtrusive.

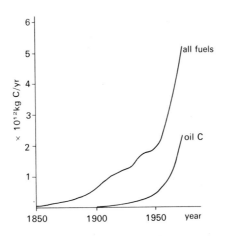

Figure 1.4 The amounts of carbon burnt per year, from oil alone and from all fuels, for the period 1850–1975.

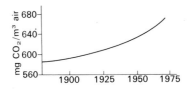

Figure 1.5 The increase in mean concentration of carbon dioxide in the atmosphere between 1880 and 1975.

The carbon cycle

An outline of the carbon cycle is shown in figure 1.3. The numbers given are estimates, based on recent chemical analyses, and are given in kg C/m² earth's surface. To obtain the actual masses of carbon in the reservoirs, or the annual turnover rates from the figures attached to arrows, the numbers should be multiplied by the surface area of the earth in m^2, which is 5.1 $\times$ 10^{14}. Study of figure 1.3 suggests that less than half of the earth's carbon is in oxidized forms such as carbon dioxide or carbonates, and rather more than half is in reduced forms. The vital reduction reaction involves photosynthesis by green plants, which can reduce atmospheric or dissolved carbon dioxide. According to figure 1.3, the residence time of carbon dioxide in the atmosphere should be 1.34/0. 274 = 4.9 years.

The uptake of carbon dioxide into the biosphere during photosynthesis is almost exactly balanced by respiration of plants, animals and micro-organisms. Small amounts of carbon are converted to coal or oil, and somewhat larger amounts are shunted back to the atmosphere as carbon monoxide or methane. Although the estimated rate of production of coal and oil is only 13 mg C/m^{2}yr, photosynthesis probably started about 2 $\times$ 10^9 years ago, so this rate could have produced 26 000 kg C/m^2 in the earth's crust. The observed figure of about 13 000 kg C/m^2 supports this rough estimate. It is important to realize that although the biosphere contains only about 0.002% of the carbon in the earth's crust, at least 95% of the remaining carbon has passed through the biological cycle, and much of it has been recycled many times.

Two effects of human activities on the carbon cycle are represented by arrows in figure 1.3. The production of calcium oxide, mostly for cement, involves the release of gaseous carbon dioxide, but the amount is less than the estimated global rate of production of calcium carbonate by corals, molluscs and other organisms. A much more marked effect on the cycle of carbon dioxide is caused by the combustion of coal and oil for the generation of energy and heat. Figure 1.4 shows the amounts of coal and oil burnt during the last century; coal is assumed to be 75% carbon and oil 85% carbon, and allowance has been made for the burning of natural gas. Figure 1.5 shows the rise in concentration of carbon dioxide in the atmosphere since 1880. This is a classic example of a perturbation of a natural cycle leading to an increase in concentration in an environmental reservoir. The total mass of carbon dioxide in the atmosphere has risen by 14% between 1880 and 1960 because we have been burning coal and oil at rates increasing from 40 to 450 times their natural rate of production during that period. In order to return to equilibrium we would have to reduce our current rate of use of coal and oil by a factor of 500, that is, to use no more than was used in the eighteenth century.

Potential consequences of this perturbation have been discussed by many authors. As far as plants are concerned, atmospheric carbon dioxide is an important plant food. The increases in its concentration may be beneficial, though in practice the amount of carbon dioxide in the air is rarely a limiting factor in plant growth. Only when temperature and light intensity are not limiting factors (as in some soft-water lakes and even in some glasshouses in temperate regions) should small increases in plant yields be expected. The extra amount in the atmosphere is far too little to cause adverse effects such as closure of stomata. As far as animal life is concerned, the small increase in carbon dioxide would be expected to have negligible effects, as the gas is scarcely toxic in the presence of a large excess of oxygen.

Meteorologists have been worried by computer predictions of the consequences of continued increases of concentrations of carbon dioxide during the next few decades. These predictions are based on the fact that carbon dioxide strongly absorbs some of the infra-red radiation from the sun. Hence an increase in carbon dioxide should increase the mean surface temperature of the earth, which in turn might melt the polar ice-caps and raise the level of the ocean. Between 1900 and 1940 the mean surface temperature of the earth did indeed rise by 0.6 K. However since 1940 the surface temperature has fallen sharply, while the carbon dioxide concentration has continued to rise. It is now believed that the main factor affecting the earth's surface temperature is the amount of dust in the stratosphere, which is replenished by major volcanic eruptions, and that the added effect of carbon dioxide is less than 30% of the effect due to dust. It is not unreasonable to predict that the concentration of carbon dioxide will reach 1.5 times its 1880 value by the year 2010, and that this may raise the mean surface temperature by about 1K.

The conclusion must be that increased inputs of carbon dioxide have caused measurable perturbations in its natural cycle, but that these do not appear to be of immediate concern. We should be much more concerned about the rates at which we are using up the exploitable resources of coal and oil, when we do not fully understand how these materials are formed and have not identified those regions of the earth where they are currently being laid down. Despite the enormous reserves of coal and oil shown in figure 1.3, most of these are either buried so deep, or dispersed in such low concentrations, that they are not commercially exploitable. Estimates of "recoverable" reserves of coal, and "proved" reserves of oil for the whole world are given in Table 1.3, together with exploitation rates in 1964. Although the estimated reserves have increased with further prospecting, the estimated residence times of these reserves is very short, especially for

Table 1.3   Estimates of the earth's exploitable coal and oil in 1964

Material	Mass $(10^{12}$kg C)	Production $(10^{12}$kg C/yr)	Residence time (yr)
Coal + lignite	1750	2	875
Oil	41.5	1·26	33

oil. The predictable drying-up of all known oil wells is already having serious economic effects on the economies of most countries.

The other aspect of the carbon cycle that deserves further study is the effect of man's exploitation of the biosphere for agricultural purposes. Estimates of the productivities of several communities are shown in Table 1.4, in terms of the mass of carbon fixed per unit area per year; in many cases these figures are based on rather few observations.

Table 1.4   Areas and mean productivities of some major communities

Community	Area (10^{14}m^2)	Productivity $(\text{kg C m}^{-2}\text{yr}^{-1})$
Forest	0.44	0.54 – 2.3
Grassland	0.27	0.91
Arable land	0.23	0.91
Desert + tundra	0.44	0.045
Ocean	3.61	0.081

From the figures in Table 1.4 we can see that changes from arable land to grassland and vice versa should have no effect on the amount of carbon fixed per year, but the conversion of forest to arable, or grassland to desert, will reduce the amount of carbon fixed. Data like those of Table 1.4 have been used to calculate the total amount of carbon fixed by vegetation, giving 1.35×10^{14} kg C/year. If all remaining forests were chopped down and converted to arable or grassland, an extremely unlikely event, the annual amount of carbon fixed would fall to 1.17×10^{14} kg C. Alternatively if all the grassland in the world were converted to desert by overgrazing, the amount of carbon fixed would fall to 1.11×10^{14} kg C. We conclude that human exploitation of the terrestrial biosphere is unlikely to affect the natural rate of fixation of carbon dioxide by more than about 20%. In the case of the ocean, where the effects of man's activities are less obvious than on land, some pessimists have predicted major losses in carbon fixation from the action of globally distributed poisons such as lead or DDT. In practice, such a disaster would have serious repercussions for fisheries, but would not reduce global fixation by more than about 22%. Both the poisons mentioned have residence times in the upper ocean of the order of tens of years.

Two further features of the carbon cycle deserve mention. The carbon monoxide shunt has been shown to have several features of interest, but is not fully understood. It is likely that plankton in the sea produce carbon monoxide, while microbes in the soil absorb the gas, on a net basis, but the rates of production and absorption are uncertain. Some carbon monoxide is oxidized to carbon dioxide in the stratosphere. Current estimates are that plankton produce about 1.75×10^{12} kg of carbon monoxide per year, to which must be added 0.28×10^{12}kg CO/yr from human and industrial activities. The internal-combustion engine can give rise to local concentrations of up to 270mg CO/m^3 in the atmosphere of some canyon-like streets of big cities. One result has been legislation in the United States to reduce the emission of carbon monoxide from car exhausts. The marked drop in carbon monoxide concentration in the atmosphere when crossing the equator from north to south probably reflects the greater number of vehicles in the northern hemisphere.

Radioactive carbon-14, with a half-life of 5760 years, has a natural concentration in recent carbon of 0.245 disintegrations per second per g C. This concentration was perturbed by nuclear-bomb tests in the 1960s, as the nuclide is produced by the reaction of energetic neutrons with atmospheric nitrogen. The perturbation, amounting to 14% of the ^{14}C in atmospheric carbon dioxide, has been useful for measuring the residence times of carbon dioxide in various environmental reservoirs, but has created problems for laboratories engaged in dating archaeological materials by the carbon-14 technique. Its long-term effects could include an increase in the mutation rate, but measurements of such an increase in human populations are not possible; for example, "natural" mutation rates are imprecisely known.

The oxygen cycle

The oxygen cycle is exceedingly complex because of the numerous chemical forms of oxygen in nature. Oxygen constitutes 46.6% of the earth's crust by weight, and a considerably greater proportion by volume. Unfortunately there are no long-lived radioactive tracers for oxygen, so that its cycle is difficult to study.

Green plants are able to oxidize water to oxygen gas by utilizing the energy of solar radiation. They are estimated to produce 260g of oxygen gas per m^2 of the earth's surface per year, and an equal amount is consumed by the respiration of bacteria and animals. At present the amount of oxygen required to burn coal and oil is about 10g/m^2 yr, and this has had no measurable effect on the concentration of oxygen in the atmosphere. Even without biological replacement, this rate of burning would require 230 000 years to use up all the atmospheric oxygen. Direct photochemical dissociation of water in the upper atmosphere is thought to add less than 1.5 g O_2/m^2 yr to the atmosphere.

The presence of large amounts of dioxygen in the atmosphere makes the earth unique among planets in the solar system. Dioxygen is thermodynamically unstable in the presence of dinitrogen and water, but fortunately its chemical reaction with these materials is negligibly slow at room temperature. Reactions which take place at slow but measurable rates at existing temperatures include the oxidations of hydrogen to water, carbon monoxide to carbon dioxide, sulphides to sulphates, and iron (II) to iron (III) salts. The relative importance of these reactions in nature, and their quantitative significance compared with respiration, is inadequately known.

The residence time for oxygen in the atmosphere is about 9000 years, and it seems very unlikely that the lack of this constituent will limit the activities of the biosphere for a period of time much greater than this. One of the least of our worries for any projection of the future is that our planet will run out of oxygen.

Of particular interest is the concentration of ozone, the unstable allotrope of oxygen, in the atmosphere. Ozone in the stratosphere is believed to be formed mainly by two chemical reactions:

(1). The dissociation of dioxygen by energetic ultra-violet radiation.

$$O_2 \longrightarrow 2O$$

(2). The addition of an oxygen atom to dioxygen

$$O + O_2 + X \longrightarrow O_3 + X$$

where X is any molecule, whose presence is needed to remove excess energy. The total amount of ozone in the atmosphere is estimated to be 6.6×10^{11} kg, and its rate of production is 22×10^{11} kg/yr, so its residence time is about 0.3 year. Ozone is an important constituent of the stratosphere because it absorbs large amounts of energetic ultra-violet radiation from the sun. If this radiation were not absorbed, it could seriously harm many terrestrial organisms. Aquatic organisms would be little affected, since water also absorbs ultra-violet radiation of wavelength less than 190nm. There is therefore some concern that the ozone layer in the stratosphere should be maintained intact.

Unfortunately the effects of introducing various pollutants into the stratosphere cannot readily be predicted. It is believed that the rate-determining step in the natural destruction of ozone is the reaction

$$O_3 \longrightarrow O_2 + O$$

which is caused by ultra-violet, visible and infra-red radiation of wavelength less than $1.18 \mu m$. The following pollutants would be expected to increase the rate of destruction of ozone: carbon monoxide, sulphur dioxide, nitric oxide, ethene and halogen atoms, as well as particulate matter, all of which are readily oxidized by ozone. On the other hand some pollutants might increase the rate of formation of ozone, for example nitrogen dioxide. Concern that aircraft flying in the stratosphere may disturb the equilibrium in the ozone layer is well warranted, but their actual effects are uncertain. It has also been suggested that fluorocarbon gases from aerosol sprays could affect the ozone equilibrium by generating stable radicals such as CF_2 or halogen atoms, but this has not been proved.

Whatever the effects on the stratosphere, it is certain that ozone can also be produced in the lower atmosphere, both by electrical discharges and by a series of photochemical reactions involving oxygen, nitrogen dioxide and alkenes in sunlight. Concentrations of up to 0.15 mg O_3/m^3 have been found locally in the troposphere, notably in Los Angeles. Such concentrations may have marked effects, notably on many growing plants, the human respiratory tract, the rate of deterioration of rubber and perhaps also of asphalt. Considerably greater efforts need to be made to understand the cycle of ozone in both the upper and lower parts of the atmosphere.

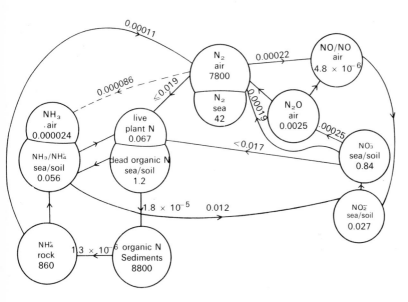

Figure 1.6 Outline of the nitrogen cycle (see caption of figure 1.3 for conventions used).

The nitrogen cycle

A simplified diagram of the nitrogen cycle is given in figure 1.6. Unfortunately there are no long-lived radioisotopes of nitrogen, so it is more difficult to prove that dynamic equilibria exist and to obtain actual rates of individual steps in the cycle. Information from the fractionation of the stable isotopes ^{14}N and ^{15}N has not been of much help in understanding the nitrogen cycle. The main reservoirs of nitrogen are the atmosphere and sedimentary rocks, which together account for 95% of all the crustal nitrogen. This suggests that virtually all this nitrogen, like crustal carbon, has been involved in the cycle shown in figure 1.6.

The important chemical transformations of nitrogen and their perturbation by human activities are listed below.

Reduction of dinitrogen to ammonia

This is accomplished in two stages in the biosphere. A large number of micro-organisms, notably all blue-green algae and some free-living bacteria, but especially the symbiotic bacteria such as *Rhizobium* spp. which live in root nodules of plants from the legume family, can convert dinitrogen to organic nitrogen of living matter. When the living matter decays, much of the nitrogen is turned to ammonia by the action of bacteria and fungi, though much is also recycled, and a small amount is retained in soil humus. The rate of conversion of dinitrogen to organic nitrogen by leguminous crops is of the order of 10 g N/m² yr, but the mean global rate is unknown. We can get a maximum figure from the amount of carbon respired per year, assuming that dry living matter contains 45% C and 3.2% N by weight. This gives a value of $274 \times 3.2/45 = 19$ g N/m² year for the whole earth, but the proportion of this which is converted to ammonia is not known.

The human perturbation of this step is the industrial fixation of atmospheric nitrogen by the Haber process. About 0.086 g N/m² yr, or 44

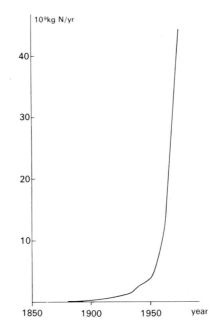

Figure 1.7 Annual amounts of atmospheric nitrogen converted to ammonia by industry.

$\times 10^9$ kg N/yr, is currently fixed, which is about 5% of the biological rate according to one estimate. Figure 1.7 shows the exponential rate of increase of industrial nitrogen fixation. Most of the ammonia produced industrially is added to the soil as fertilizer, either directly or after oxidation to nitrates.

Oxidation of ammonia to nitrate

This takes place in two stages, both in soils and the ocean. The bacterium *Nitrosomonas* oxidizes ammonia to nitrite, and a second species *Nitrobacter* converts nitrite to nitrate. The rate of conversion is probably of the order of 10 g N/m² yr.

Industrial production of nitrates for fertilizers is about 7×10^9 kg N/yr, or 0.014 g N/m² yr assuming uniform distribution over the whole planet. In practice, the application of nitrate fertilizers is not at all uniform, with the largest amounts being used in Western Europe and North America. In some parts of the central United States, and in California, nitrates applied to the soil have subsequently appeared in groundwater feeding local wells. Nitrates themselves are not very poisonous, but they can be readily reduced to the much more toxic nitrites by intestinal bacteria. Well waters containing more than 10μg nitrate/l are not recommended for human consumption, as they could damage the health of infants; greater concentrations than this have occurred from excessive use of fertilizer in the United States. Nitrates are also a contributory cause of eutrophication in lakes or reservoirs, which will be discussed below

Ammonia and nitrate behave very differently in soils. The ammonium ion is strongly retained by clays, is immobile, and has a long residence time. On the other hand the nitrate ion is not absorbed by soils, but is extremely mobile, moving both horizontally and vertically with soil water. There is some reason to believe that it is better for the environment if applied fertilizers contain ammonia rather than nitrate nitrogen, but interconversion of the two forms can be rapid. Certainly nitrogen is excreted and recycled in the reduced form in natural waters, and added nitrates are only used when other sources of nitrogen have been depleted.

Conversion of nitrates to dinitrogen

Many bacteria are able to convert nitrates to gaseous nitrogen, and indeed many organisms are able to use the oxygen from nitrate ions instead of atmospheric oxygen, when the latter is in short supply. One estimate of the natural rate of this process is 0.16 g N/m² yr, but the value is uncertain. It is virtually unaffected by human activities except in situations where excessive amounts of nitrates are used as fertilizers.

Production of oxides of nitrogen

The production of nitrous oxide (N_2O) is a side-shunt in the main nitrogen cycle. The gas is produced by microbes both in the soil and in the sea, but whether oxidation of organic nitrogen or reduction of nitrates is the more important process is uncertain. The rate of production is estimated to be about 0.4 g N/m^2 yr, giving a residence time in the atmosphere of about 10 years. The gas is readily absorbed by the soil, but some may be destroyed in the stratosphere by ultra-violet radiation, giving rise to nitric oxide as well as nitrogen and oxygen atoms.

$$N_2O \begin{cases} \longrightarrow N_2 + O \\ \longrightarrow N + NO \end{cases}$$

The nitrous oxide subcycle does not seem to be affected by man's activities, and the gas is not known to have any effects on organisms at its natural concentrations.

Nitric oxide (NO) is a more reactive gas, with an unpaired electron, and is produced by several processes, e.g.

1. $\qquad NH_3 \xrightarrow{\text{microbes}} NO \qquad$ in both soil and sea

2, 3 $\qquad N_2 + O_2 \longrightarrow 2NO \qquad$ in the atmosphere, by (2) combustion or (3) lightning

4. $\qquad N_2O \xrightarrow{\text{radiation}} NO + N \qquad$ in the stratosphere

Quantitatively, reactions 1 and 2 are believed to be much more important than 3 and 4, and it is thought that inputs from combustion only compete with microbial inputs locally, near major centres of industrial activity. The rate of production of nitric oxide during combustion in air depends critically on temperature. Its residence time in air is not known, but is probably in the range 7–70 days, as oxidation to nitrogen dioxide is surprisingly slow at natural concentrations. It is not very harmful to plants and animals, but is a source of several much more toxic products, such as nitrogen dioxide, peroxoacetyl nitrate and the type of smog prevalent in Los Angeles.

Nitrogen dioxide (NO_2) is a still more reactive brown gas produced when nitric oxide and oxygen react together.

$$2NO + O_2 \longrightarrow 2NO_2$$

ts residence time in the atmosphere is of the order of three days, because of its chemical reactivity and high solubility in water. It is probably responsible for the brown-coloured air which can be seen when looking down the length of traffic arteries in big cities. The gas rarely exists in the atmosphere outside city centres at concentrations large enough to damage plants or animals by direct uptake. It is, however, clearly implicated in the production of more toxic substances such as ozone and peroxoacetyl nitrate (in the presence of hydrocarbons). For this reason, legislation has been enacted in the United States to reduce the emission of both nitric oxide and nitrogen dioxide from vehicle exhausts, which can be done by inserting suitable catalysts in the exhaust manifold. Air pollution by oxides of nitrogen is a local problem that can be controlled by applying existing technology. It is estimated that only about 14% of the nitrogen dioxide in the atmosphere is contributed by combustion processes, the remainder being produced by bacteria.

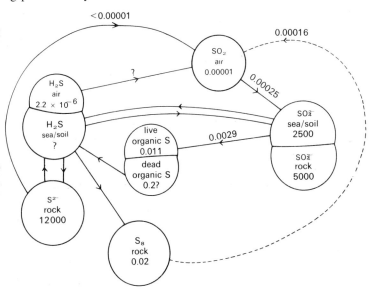

Figure 1.8 Outline of the sulphur cycle (see caption of figure 1.3 for conventions used).

The sulphur cycle

The sulphur cycle is outlined in figure 1.8. There are enormous amounts of both sulphates and sulphides in the earth's crust, together with large amounts of sulphate ion in the ocean. The main uncertainties in the sulphur cycle are the quantities of free hydrogen sulphide in the atmosphere, and the rates of nearly all the processes involved. Most of the oxidations and reductions in the sulphur cycle are mediated by common species of bacteria. Green plants obtain the bulk of their sulphur from the soil as sulphates, which they are able to reduce to organic sulphides or hydrosulphides. Animals are not able to reduce sulphates.

There are two major perturbations of the sulphur cycle by man, the production of sulphur dioxide and the mining of elementary sulphur. The production of large amounts of sulphur dioxide and its injection into the atmosphere occurs as a by-product of combustion. Most heavy fuel oils and commonly used coals contain on average 1.5% sulphur, and much of this escapes from chimneys as sulphur dioxide when the fuel is burnt. Some North American coals contain greater amounts of sulphur; this restricts their use. In addition to the combustion of fuel, the smelting of sulphide ores, especially those of copper, lead and zinc, probably contributes about 10% of the total annual emission of sulphur dioxide. The only known natural source of the gas is volcanic emission, estimated to produce $2—5 \times 10^9$ kg S/year, which is much less than the 83×10^9 kg S from industrial and domestic sources in 1975.

The residence time of sulphur dioxide in the atmosphere is still uncertain, but is probably about 4 days. The uncertainty arises because the gas is removed from the atmosphere by several different processes, whose rates are of the same order of magnitude, e.g.

1. Wash-out in rain, either as SO_2 or sulphates (75%).
2. Absorption by plants, soils or the ocean (25%).
3. Chemical oxidation.

The short residence time means that sulphur dioxide is both a local and a seasonal pollutant. Concentrations in the atmosphere are very variable, but on average they are highest above or to leeward of big cities (figure 1.9). They are also higher in winter, when more fuel is burnt, than in summer (figure 1.10).

The annual amount of industrial sulphur injected into the atmosphere (83×10^9 kg S) and the amount falling back to the surface of the earth (212×10^9 kg S) do not balance. Part of the difference comes from the 44×10^9 kg S/yr which is estimated to be entrained as sulphates in fine particles of ocean spray, and the remainder (85×10^9 kg S/yr) is assumed to be hydrogen sulphide and/or volatile organic sulphides or thiols produced by microbial action on dead organic matter or excreta. There is a great need to confirm this assumption by actual measurements of volatile sulphides in the atmosphere. Since such sulphides probably have short residence times, of the order of days, their concentrations probably vary with altitude, season and location. It is supposed that man's activities have not perturbed the hydrogen sulphide subcycle to anything like the extent that they have altered the sulphur dioxide subcycle.

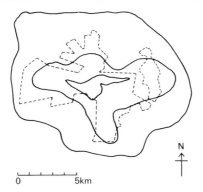

Figure 1.9 Mean sulphur dioxide contours in the air around Reading, Berks (population 120 000). The dotted line encloses the built-up area. The three concentric contour lines represent concentrations of 30, 60 and 90μg SO_2/m³ air.

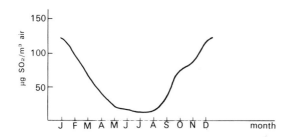

Figure 1.10 Seasonal variation in sulphur dioxide concentration at a single recording station in Reading, Berks (4-year average).

Sulphur dioxide is a serious local pollutant with a wide range of known effects. Its only beneficial effects are supposed to be the relief of sulphur deficiencies in soils, which are rare in Britain, and the control of certain fungal diseases, such as "black spot" in roses *(Diplocarpon rosae)*. Demonstrably harmful effects include:

1. The production of "acid rain" from its oxidation product, sulphuric acid, which leaches nutrients from soils in the United States and Scandinavia.
2. The erosion of limestone and mortar in city buildings, caused by conversion of calcium carbonate to the more soluble sulphate.
3. The embrittlement of paper, especially in books stored for long periods in city libraries. Sulphur dioxide is readily absorbed by paper, and oxidizes to sulphuric acid which attacks cellulose.

D

4. The corrosion of metals exposed to city atmospheres, especially aluminium, copper, iron and zinc, but not lead, whose sulphate is relatively insoluble.
5. The extinction of many species of lichens which grow on bark, over large areas. Thus Holland has lost 27% of its bark lichen flora during the last century, and England has suffered similar losses in the industrialized axis between London and Liverpool (figures 1.11 and 1.12). It appears that about 50 species of lichen are quantitative indicators of the intensity of sulphur dioxide pollution. Pollution maps based on these indicator species are cheaper to produce than are maps based on automated gas analysis. The minimum concentration of sulphur dioxide causing damage to lichens is $30\mu g/m^3$, which is about ten times the lowest background concentrations reported in the few remaining unpolluted parts of the world.

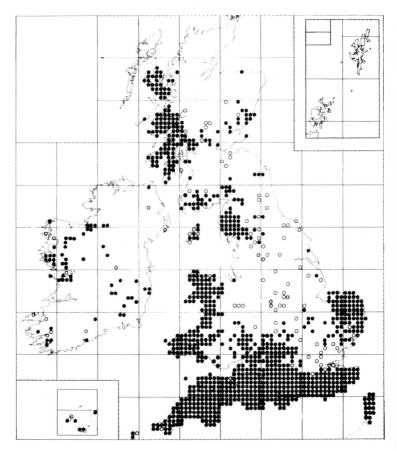

Figure 1.11 Distribution of the lichen *Parmelia caperata* in the British Isles, based on data from the British Lichen Society's Distribution Maps Scheme. ● = post–1960 records, ○ = old records. Note absence or extinction from regions of industrial activity.

6. The reduction in yield of green plants. The gas produces a characteristic necrosis of interveinal areas of the leaves, which can be seen in many suburban hedges. Species vary greatly in sensitivity to sulphur dioxide, but none are as sensitive as lichens, since no effects have been reported when concentrations of the gas are below $100\mu g/m^3$. In general, evergreen species are more sensitive than deciduous plants, since they have to cope with higher winter concentrations of sulphur dioxide. For example, conifers can no longer be established on that part of the western slopes of the Pennine Chain which receives pollution from Liverpool and Manchester. Fortunately most agricultural crops are grown during the summer, when sulphur dioxide concetrations are low.

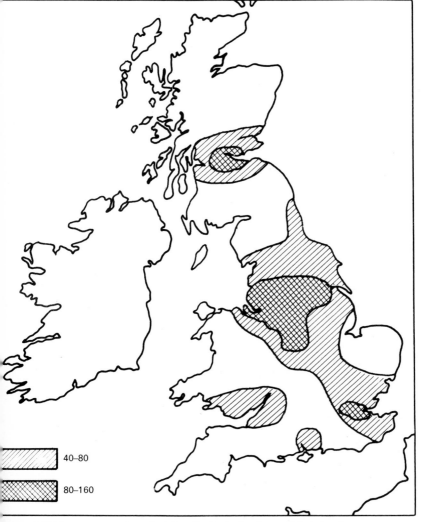

40–80

80–160

Figure 1.12 Mean concentration of sulphur dioxide in the air over Britain in micrograms per cubic metre. Compare with figure 1.11.

7. The irritation of the human respiratory tract. Sulphur dioxide is not poisonous to man even at concentrations where it can just be detected by smell, but chronically high concentrations can cause respiratory problems. The excess mortality caused by city smogs during inversions is mainly due to particulate matter, with sulphur dioxide exerting a minor effect.

In view of all these problems caused by sulphur dioxide, it is not surprising that attempts have been made to restrict the release of the gas. Some advocate removal of sulphur from the fuel, while others would prefer to remove sulphur dioxide from the combustion gases. It is technically possible, though expensive, to remove sulphur from heavy fuel oils before burning them. It is not technically possible to remove sulphur and/or pyrites from coal on a large scale. A number of processes have been tested in pilot plants for removing sulphur dioxide from industrial combustion gases, but there is no method suitable for removing the gas emitted from domestic chimneys. The cost of removal in large installations is reasonably low, and may be offset by the value of any sulphur recovered. For example, in Britain as much sulphur goes up our chimneys (as waste sulphur dioxide) as the total amount in sulphuric acid used by industry, so the recovery of this sulphur would be a worthwhile saving. At present British policy is to build taller chimneys, so that the sulphur dioxide is more widely distributed and diluted.

The other perturbation of the sulphur cycle by man is the consumption of sulphur deposits from underground salt domes. Such deposits, of which about 2×10^{11} kg are estimated to exist in the south-east United States, are believed to have been formed by anaerobic bacteria (*Beggiotoa* spp.) from hydrogen sulphide. If we suppose that such sulphur deposits occur, as yet undiscovered, as densely in the rest of the world as in the United States, this gives an estimate of about 10^{13} kg of "bacterial" sulphur as a global resource. If all this has been formed in the last 500 million years, which is geologically reasonable, its rate of production would be 2×10^4 kg/year. The rate of exploitation of this sulphur resource by the Frasch process was about 10^{10} kg/year in 1964, and it is not surprising that the American deposits are already nearing exhaustion. The parallel with the world's drying oil wells is a close one.

The phosphorus cycle

The phosphorus cycle is simpler than the cycles of carbon, nitrogen and sulphur for several reasons. Firstly, the element is very rarely oxidized or reduced during its cycle, but remains in the pentavalent oxidation state (reduction to phosphine (PH_3) can occur in lake sediments, but in negligibly small quantities). Secondly, natural organic phosphates are all derived from phosphoric acid, and the phosphorus-carbon bond is not found in nature. Thirdly, there are no volatile phosphates, so relatively little phosphorus cycles through the atmosphere. Elemental phosphorus, as the P_4 molecule, is an extremely poisonous substance, but is absent from natural sources and is very rarely found as a pollutant. The phosphorus cycle is concerned with the mobilization and transfer of the phosphate ion between various reservoirs, as outlined in figure 1.13. Minor reservoirs of phosphates, such as living organisms, fresh waters and their sediments, and guano deposits have been left out in figure 1.13.

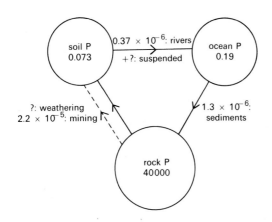

Figure 1.13 Outline of the phosphorus cycle (see caption of figure 1.3 for conventions used).

Our knowledge of the rates of transfer of phosphate between reservoirs is limited, but the best estimate is probably the sedimentation rate of 6.4×10^8 kg P/yr, corresponding to a residence time in the ocean of about 160 000 years. The rate of transfer of soluble phosphates from land to sea by rivers is estimated to be about 1.9×10^8 kg P/yr, but it is probable that about three times as much is carried down in suspension as in true solution. It is likely that until about 1900 the phosphate cycle was in true equilibrium, with the rate of rock weathering equal to the rate of river transfer equal to the rate of sedimentation. This would imply a residence time of 56 000 years for phosphorus in soil.

Human exploitation of rock phosphates has been increasing exponentially this century (figure 1.14), and was estimated as 1.12×10^{10} kg P in 1975. Production overtook the natural rate of cycling about 1910. Between 75 and 90% of rock phosphates are ultimately applied to agricultural soils as fertilizers, either ground up or as more soluble processed products. Much of the remainder is used in water softening, for

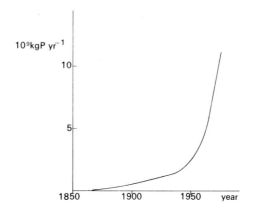

Figure 1.14 Annual amounts of phosphates used in agriculture and industry.

example as sodium tripolyphosphate, and this is discharged directly to rivers.

Phosphates in solution are rapidly converted to sparingly soluble forms. The phosphates of aluminium, calcium and iron have notably low solubilities, which depend on pH, and trace quantities of phosphates are strongly absorbed by hydrous oxides of iron or aluminium. This low solubility accounts for the long residence time in soil, and for the low concentration of phosphate found in sea water and most freshwater systems, including soil solutions. One consequence is that soluble phosphate is often, though not always, the nutrient which limits plant growth in soils, fresh waters and the ocean, especially in infertile soils, lakes and coastal waters. For example, the phosphate concentration in the upper layer of the ocean is very much less than in the deeper layers, because of its uptake and rapid recycling by plankton near the surface. Hence it is not surprising that the addition of large amounts of phosphates to agricultural soils has increased crop yields. Some of this added phosphate is lost in drainage water.

The injection of phosphates into freshwater systems, as domestic, industrial or agricultural wastes, often results in excessive growth of algae which are of no commercial use. The process of injecting plant nutrients is known as "eutrophication", and has important biological consequences for natural lakes and reservoirs. These include phenomenal blooms of algae, which make the water turbid or unsightly, followed by their decay, which can render the water locally anaerobic and cause bad tastes and smells. Eutrophication is also associated with the replacement of game fish, such as salmon and trout, by coarse fish, and with an increased production of organic matter, insects and other invertebrates. Flowing systems such as rivers purify themselves quickly, but in large lakes with residence times for water of the order of tens or hundreds of years, the effects of eutrophication appear slowly. In river basins without lakes, added phosphates will mostly end up in estuarine muds, which are highly eutrophic. Eutrophication is known in freshwater lakes all over the world, notably in the Great Lakes in North America and the Zürichsee in Switzerland. One of the best methods of controlling it has been found to be the removal of phosphates from the input waters. This is technically possible by treating contaminated input streams with aluminium sulphate, usually in commercial sewage plants. The phosphate is precipitated as aluminium phosphate, which can theoretically be recycled.

The fluorine cycle

The fluorine cycle is little understood. Volcanoes are known to emit small and variable amounts of hydrogen fluoride, a very poisonous and corrosive gas with a residence time in air of a few days. The rate of emission may be of the order of 10^{10} kg F/year. Hydrogen fluoride is very soluble in water, and is readily absorbed by soils, where it may react with limestone to form the insoluble calcium fluoride or become fixed in clay minerals. It is believed to have a long residence time, of the order of 10^4 years, in soils. Fluorides carried into the sea by rivers are co-precipitated with calcium phosphate and calcium carbonate; the residence time of the fluoride ion in the ocean is about 900 000 years. Phosphatic sediments are rich in fluorine, so that phosphate fertilizers enrich the element in cultivated soils.

Hydrogen fluoride is also emitted locally by some industrial operations, especially by brickworks, glassworks, aluminium factories and some ironworks. Nearby trees are affected and tend to become stag-headed, but some of the damage may be caused by sulphur dioxide. The gas kills lichens even more effectively than does sulphur dioxide, and erodes siliceous building materials such as brick and sandstone. Sometimes the amounts emitted are sufficient to affect animals grazing the local vegetation. Cows are much more sensitive than sheep, and have suffered severe bone damage near some of the Bedfordshire brick kilns. The hazard is a very local one, and has been solved by abandoning cattle raising in the affected fields. The gas has been suggested as a contributory cause of lung cancer in Hamilton, Ontario.

The removal of hydrogen fluoride from effluent gases should be possible from a technical point of view, but would add to the costs of the industries involved. There are no available estimates for the amounts of hydrogen fluoride emitted annually by industrial processes.

Heavy metal cycles

Copper, lead and zinc

These three metals occur chiefly as sulphides in nature. In their natural cycle, the sulphides are partly oxidized to more soluble compounds such as sulphates by weathering. The metal ions are carried down to the sea by rivers, where they are fairly rapidly removed by precipitation, mostly as sulphides. Lead may also be precipitated as a complex phosphate, and all three metals may be co-precipitated with hydrous manganese oxides or calcium carbonate. Estimates of the natural rates of cycling and residence times in the ocean are given in Table 1.5.

Table 1.5 Major cycles of copper, lead and zinc

Metal	*River transfer* $(10^9 kg\ yr^{-1})$	*Loss to sediments* $(10^9 kg\ yr^{-1})$	*Residence time in sea* (years)
Cu	0.20	0.31	2000—4000
Pb	0.15	0.05	300—850
Zn	0.37	0.26	19000—27000

The figures given in Table 1.5 are only estimates and may be wrong by a factor of two. It seems that river transfer and loss to sediments are roughly equal for copper and zinc, but not for lead.

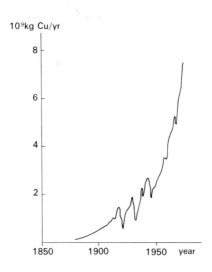

Figure 1.15 Annual industrial production of copper.

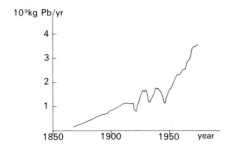

Figure 1.16 Annual industrial production of lead. Note change in slope around 1940, when lead alkyls were first used as petrol additives.

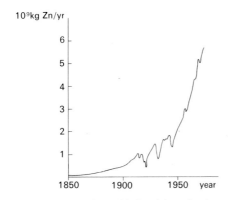

Figure 1.17 Annual industrial production of zinc.

The rates of extracting minerals of these three metals during the last hundred and twenty years are shown in figures 1.15–1.17. Much of the metal produced is exposed to the weather. All three metals are used for roofing materials, galvanized wire is used for fences, and copper wire for transmitting electricity. Atmospheric corrosion converts them all to soluble salts, especially in industrial areas where there is much sulphur dioxide, and returns them to the weathering cycle. Under natural conditions, zinc corrodes faster than copper, while lead corrodes very slowly.

All three metals are added to the atmosphere in the smoke from burning coal, but the amounts are quite small, probably about 5–10×10^5 kg/yr of each metal over the whole globe. The lead contribution is dwarfed by the amounts of lead injected into the atmosphere from burning petrol in cars. Since the 1940s, lead alkyls have been added to petrols as anti-knock agents, and currently about 10% of the world production of lead is used for this purpose. Hence about 4×10^8 kg of lead was added to the atmosphere in 1975. It is not surprising that lead aerosols are now global contaminants. Such aerosols have a residence time in air of about a month, and have caused measurable increases in the concentration of lead in the upper layers of soils, peat deposits, lake sediments, North Greenland ice, many parts of the ocean and recent tree rings. Zinc and copper have also increased in recent marine and lake sediments near industrial regions (figure 1.18).

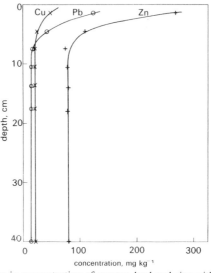

Figure 1.18 Variation in concentration of copper, lead and zinc with depth of sediments in Lake Michigan. The rate of sedimentation is about 1 mm/yr, implying a change about 80 years ago.

From a study of soils, it is doubtful if any part of the world remains uncontaminated by lead. In industrialized countries the input of lead from rainfall and dry dustfall greatly exceeds the output (Table 1.6). Table 1.6 shows that widespread contamination of soils by atmospheric copper and zinc also occurs, but is much less of a problem, since these two elements are both essential to plants and are regularly removed by cropping. On the other hand lead is rapidly accumulating in the upper layer of topsoil in central England and elsewhere. This overall increase in soil lead has not yet been shown to have any measurable effects on the welfare of living plants and animals. Recently there has been concern that some behavioural disorders of city children may be symptoms of lead intoxication, but this is not attributed to soil-borne lead. Local contamination by copper, lead and zinc has occurred near ore smelting works, and when mine wastes, certain sewage sludges and some estuarine muds have been spread on soils. The consequences of such local pollution can be serious and very long-lived, e.g.

he death of livestock, and killing or stunting of plants near smelters, the
ailure of vegetation to colonize mine wastes, and the stunting of cereals
grown on zinc-rich muds from the Thames estuary.

Table 1.6 Estimated inputs and outputs to Thames valley soils

Element	Input (mg m^{-2}yr^{-1}) (rain, fertilizer, etc.)	Output (mg m^{-2}yr^{-1}) (cropping, drainage)	Residence time in soil (yr)
As	1.9	0.24	2000
Cd	1.1	0.7	280
Cr	6.8	0.1	6300
Cu	11.8	6	860
Hg	0.082	0.013*	920
Ni	6	2	2300
Pb	38	1.4	3000
Zn	48	59	2100

neglecting unknown losses by volatilization

Concern about global pollution by lead, which is not known to be
essential to life and is much more toxic to mammals than either copper or
zinc, has given rise to legislation restricting the use of leaded fuels in cars in
he United States. Small reductions have been made in the amounts of lead
added to European petrols. Petrols used in the Soviet Union have always
been free from lead.

There is also cause for concern in the rate of usage of known ore bodies.
In 1800 Britain was the world's largest producer of copper, mostly from
Devon and Cornwall. These deposits are now almost entirely exhausted, as
are the lead mines in Cardigan, Lanarkshire and elsewhere. These examples
should serve as a warning of the finite nature of mineral resources. From a
global point of view, lead and zinc ores are still in fair supply, but the better
copper ores have mostly been used up, and ores containing as little as 0.4%
Cu are being processed. Increased costs, resulting from the use of such poor
quality ores, have led to more efficient recycling of copper and lead scrap.

Arsenic, cadmium, chromium, mercury and nickel

Information on the natural cycling of these metals is summarized in Table
1.7. These five elements have been chosen because industrial production
now exceeds the annual amounts in the natural cycle of water; data for
ndustrial production are given in Table 1.7 and figures 1.19–1.23.

In the case of arsenic, industrial production is small and is probably
exceeded by the amounts liberated as a by-product of burning coal. An
unknown proportion of the arsenic released by man is cycled through the

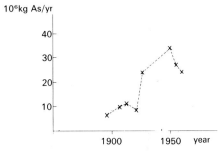

Figure 1.19 Annual industrial production of arsenic. Few figures are available. No allowance is made for arsenic released as a by-product of burning coal.

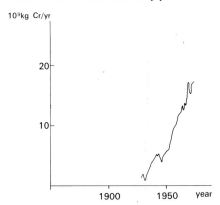

Figure 1.20 Annual industrial production of cadmium.

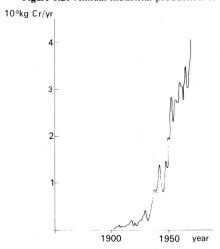

Figure 1.21 Annual industrial production of chromium. A substantial fraction of the production is used as chromite without reduction to the metal.

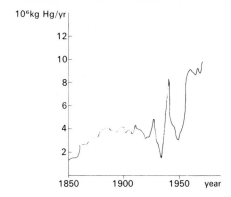

Figure 1.22 Annual industrial production of mercury. No allowance is made for mercury released as a by-product.

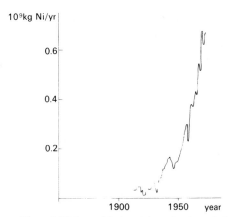

Figure 1.23 Annual industrial production of nickel.

Table 1.7 Major cycles of As, Cd, Cr, Hg and Ni

Element	River transfer (10^6kg yr)	Loss to sediments (10^6kg yr)	Residence time in sea (kyr)	Industrial production (10^6kg/yr)	Coal by-products (10^6kg/yr)
As	9.6	5.6	550–930	35	100
Cd	1.1	0.62	140–250	16	2
Cr	210	640	0.1–0.34	3000	10
Hg	3.7	1.1	12–40	9.5	1.6?
Ni	370	300	8–9	667	120

Notes: No allowance is made for elements cycled through the atmosphere. Industrial production figures for 1970. Coal by-product figures assume 2×10^{12}kg of coal were burnt in 1970, but the proportions of each metal appearing in ash, smoke and gaseous forms is uncertain.

air. The data of Table 1.6 indicate that rural soils may be accumulating arsenic in an industrialized country like Britain, while those of Table 1.7 show that the burning of coal could be responsible. In the United States lead arsenate has been applied to large areas of cultivated soil as a fungicide, but the use of this product has declined since 1970.

Arsenic is a well-known poison with no certain function in living things. It is more toxic to mammals than it is to plants, and it is probably carcinogenic. Many sea foods contain unusually large amounts of arsenic. Marine bacteria, and some terrestrial fungi, are able to methylate arsenic, using Vitamin B_{12} as a methylating agent. One product is the very poisonous trimethyl arsine, but whether such volatile forms take part in the natural cycle of arsenic through the atmosphere is unknown.

Cadmium is an industrial pollutant which has reached high concentrations in certain estuaries, for example, the Severn estuary which receives drainage waters from the old zinc-smelting works in South Wales. It does not appear to be accumulating in soils, though it is sometimes present in high concentrations in sewage sludge.

Pollution of the River Jintsu in Japan has caused serious problems to the local population (itai itai disease). The unpleasant consequences of excessive intake of cadmium by mammals include kidney damage, sterility and cancer. The metal is not implicated in damage to plant life.

The cycle of chromium is poorly understood. The annual river transfer and sediment losses do not balance, nor do the inputs and outputs of British soils; there is no obvious reason why there is such a high input to soils from rain and dust. One reason is that the analysis of chromium is difficult, so that the figure given in Tables 1.6 and 1.7 may be wrong. It is also possible that micro-organisms are able to convert chromium to volatile forms which cycle through the atmosphere. Some biological materials suffer large unexpected losses of chromium when heated.

Chromium is now thought to be an essential element, and may be deficient in some diets of refined foods, e.g. in the United States. It is not particularly poisonous, but dusts which contain chromium are associated with lung cancer. Actual pollution damage due to chromium is rare and localized to tips containing chromium-rich wastes. Soluble chromium is highly toxic to vegetation.

The cycle of mercury has recently been shown to be quite complicated, and work on it is proceeding apace. Micro-organisms are able to reduce mercury (II) salts to metallic mercury, and also to methyl mercury derivatives or even the highly toxic dimethyl mercury. All these are soluble in water at trace levels, since even metallic mercury dissolves to the extent of 20 mg/l at room temperature. They are also more or less volatile, and one recent estimate is that the amount of mercury volatilized from the earth's rust each year is about 84 × 10⁶ kg. This is much larger than the amount thought to be cycled by rivers, and also greatly exceeds the annual industrial production of mercury. It may be that large amounts of volatile forms of mercury enter the atmosphere from burning coal, but there are few reliable analyses of mercury in coal and the fossil fuel contribution is estimated to be 1.6 × 10⁶ kg/yr. Mercury is certainly concentrated by some coals, and also in the upper layers of soils. Its input to British soils appears to exceed output (Table 1.6), but no allowances have been made for gains from and/or losses to the atmosphere in this table. Further research is needed to establish the chemical nature and relative proportions of the volatile mercury species in the atmosphere, which may include mercury atoms, inorganic mercury (II) salts and organomercurials as well as mercury species absorbed on dust.

For a long period, 1880 to 1950, the industrial production of mercury (neglecting any contribution from burning coal) was about the same as the natural rate of cycling in water. After 1950 two things changed—the industrial production approximately doubled, and there was a corresponding large increase in the use of organomercurial fungicides as seed dressings. It was soon found that large increases in the mercury contents of certain birds and fishes were occurring. One result was the banning of methyl mercury fungicides from many countries; the less volatile and less toxic phenyl mercurials are used instead. Mercury concentrations in fish, marine molluscs and crustacea have always been higher than in other foods. There does not appear to be any danger of man's activities polluting the open ocean with mercury, but concern for freshwater lakes and estuaries is warranted.

All mercury salts are poisonous, particularly the fat-soluble methyl mercurials and the water-soluble mercury (II) salts. Mercury metal is less toxic because of its small solubility when swallowed, but the vapour is dangerous to inhale. There have been several local incidents of mercury pollution with lethal consequences, such as pollution of estuaries in Japan, and human consumption of seed grain treated with organomercurials in Iraq. Surprisingly, there appear to be no reports of mercury poisoning of plants or animals in the vicinity of mercury mines.

Nickel appears to be in balance as far as global cycling is concerned, but in industrial countries the input to soils may exceed output, so that soils may be accumulating nickel. This could be serious in the long term, because nickel is one of the most poisonous elements to plants, and there are large areas of land which are unsuitable for agriculture because of naturally high concentrations of nickel. These areas have soils derived from serpentine, are rich in both nickel and chromium, and have low ratios of calcium to magnesium. In nature they carry a specialized vegetation which is usually sparse and is made up of dwarf species. It is possible, though quite unproven, that some nickel cycles as the volatile and extremely poisonous tetracarbonyl, $Ni(CO)_4$.

There have been few reports of nickel toxicity to man except in factories where large amounts of metal are handled. Apart from causing dermatitis nickel dusts have been held responsible for some cases of lung cancer.

Summary. History of perturbations of natural cycles

Table 1.8 summarizes the main perturbations of natural cycles by the elements discussed in this chapter. The dates given in the last column represent approximate past or projected future times when the annual turnover by human activities has equalled or will equal the natural turnover in the particular step of the cycle specified. Future dates are predicted by assuming a continuation of current trends with no intervening catastrophes.

Table 1.8 Major human interferences with the rates of natural cycles

Element	Step of cycle	1975 ratio $R = \dfrac{\text{human use}}{\text{natural use}}$	Year when $R = 1$
C	Release of CO_2 to atmosphere	0.04	21st century
C	Consumption/natural production of coal	500	17th century
C	Release of CO to atmosphere	0.16	21st century
O	Consumption/natural production of O_2	0.04	Never?
N	Consumption/natural production of N_2	0.05	21st century
N	Production of NO_2	0.13	21st century
S	Production of SO_2	17	1885
S	Mining/microbial production of S_8	5000	19th century
P	Mining/sediment gain	17	1910
As	,, ,,	6	1910
Cd	,, ,,	29	1920
Cr	,, ,,	5	1940
Cu	,, ,,	27	1880
Hg	,, ,,	9	1880–1950
Ni	,, ,,	2	1960
Pb	,, ,,	70	1860
Zn	,, ,,	22	1880

The data of Table 1.8 suggest that the most serious human interferences have been the consumption of world stocks of coal, oil and sulphur, for which there is no global remedy. The atmospheric pollutant whose balance has been most disturbed by human activites is sulphur dioxide, which despite its short residence time has had widespread biological and corrosive effects. Metals are not irreversibly consumed like coal or oil but tend to be broadcast in dilute form by man's activities. The greatest interference has been in the cycling of lead, but the cycles of copper, cadmium and zinc have also been disturbed. Since these metals have long residence times in the environment, the ultimate effects of these perturbations remains uncertain.

FURTHER READING

Anon. (1974), *Metal Statistics 1963–1973*, 61st ed., Frankfurt.

Bowen, H. J. M. (1966), *Trace Elements in Biochemistry*, Academic Press.

Broecker, W. S. (1974), *Chemical Oceanography*, Harcourt Brace Jovanovich.

Delwiche, C. C. and Bolin, B. (1970), "The Carbon and Nitrogen Cycles", *Sci. Amer.* 223, 125 and 136.

Junge, C. E. (1963), *Air Chemistry and Radioactivity*, Academic Press.

Junge, C. E. (1972), "The Cycle of Atmospheric Gases", *Quart. J. R. Meteorological Soc.* **98,** 711.

Polunin, N. (Ed.) (1972), *The Environmental Future*, Macmillan.

Singer, S. F. (Ed.) (1970), *Global Effects of Environmental Pollution*, D. Reidel (Holland).

E

CHAPTER TWO

MERCURY

LEONARD J. GOLDWATER AND WOODHALL STOPFORD

Introduction

Quite in keeping with its elusive nature, mercury, also known as quicksilver, has had a checkered career. In 1742 the great British physician George Cheyne wrote:

> Mercury, judiciously manag'd, seems to me to be the only true Panacea and universal Antidote, sought by wise and boasted of by pyrotechnical Enthusiasts. Mercury seems pointed out and impress'd by the Signature of the God of Nature, for the Cure, at least the Relief, of intelligent Creatures made miserable by hereditary Diseases, by natural appetites irregularly indulg'd, by Ignorance and bad Example and Frailty in the human Kind, especially made so by high Food and spiritous Liquors mostly.

A hundred years earlier, Cheyne's countryman, the military surgeon John Woodall, had stated of mercury:

> It is the hottest, the coldest, a true healer, a wicked murderer, a precious medicine, and a deadly poison, a friend that can flatter and lie.

With a background such as that, it is hardly surprising that in the last quarter of the twentieth century mercury should find itself once again embroiled in controversy. Is it a precious medicine or a deadly poison? Is it a friend or a foe? Actually it is all of these, and more.

General and physical properties

Occurrence

Mercury in some form has been found everywhere it has been sought, and hence is generally considered to be ubiquitous throughout nature. This was suspected as long ago as 1934 and has been repeatedly confirmed during the past decade as a result of the development of highly sensitive analytical methods.

Table 2.1 Mercury-bearing minerals*

Name	Formula	Where Found
Arquerite	$AgHg_3$, Ag_5Hg_3, Ag_6Hg, Ag_2Hg_5	Chile
Barcenite	Antimonate of Mercury	Mexico
Bordosite	$AgHgI$, $AgCl \cdot 2HgCl$	Chile
Calomel	$HgCl$	Texas, USA, Yugoslavia, Germany, Italy
Cinnabar	HgS	All continents except Antarctica
Coccinite	Hg_2OCl (?)	Mexico
Coccinite	HgI_2	Australia
Coloradoite	$HgTe$	Colorado, USA
Eglestonite	$Hg_4 Cl_2O$	Texas, USA
Gold Amalgum	Au_2Hg_3, Au_2Hg_5	California and Oregon, USA, Borneo
Guadalcazarite	$Hg_5 \cdot Zn \cdot Se$	Mexico
Hermesite	Tetrahedrite + Hg	Bavaria, Germany
Idrialite	$HgS + C_3H_2$	Yugoslavia
Iodargyrite	$AgHgI$ (?)	Germany, Spain, France, Congo, Chile, USA, USSR, Australia
Kalgoorlite	$Ag_2Au_2HgTe_6$	Australia, Colorado
Kleinite	$Hg \cdot NH_4Cl \cdot SO_4$ (?)	Texas, USA
Kongsbergite	$AgHg$	Norway
Lehrbachite	$HgSe + PbSe$	Germany
Leviglianite	$HgS + Zn$	Italy
Livingstonite	$HgSb_4S_7$	Mexico
Magnolite	Hg_2TeO_4	Colorado, USA
Metacinnabarite	HgS (dimorphous)	Mexico
Montroydite	HgO	California, Texas, USA
Moschellandsbergite	Ag_2Hg_3	Sweden, Bavaria, Germany, France
Mosesite	$Hg_6(NH_3)_2Cl_2(SO_4)(OH)_4$	Texas, Nevada, USA
Onofrite	$ZnS \cdot 6HgS$	Oregon,.USA
Potarite	Pd_3Hg_2	British Guiana
Schwatzite	Tetrahedrite + Hg	Tyrol
Terlinguaite	Hg_2OCl	Texas, USA
Tiemannite	$HgSe$	Harz Mts., Germany, Utah, USA
Tocornalite	$(AgHg)I$	Chile

*Adapted from Dana's *The System of Mineralogy* (1944-62), with additions.

By far the greatest part of naturally-occuring mercury is in the form of the sulphide (cinnabar) but, as shown on Table 2.1, there are at least thirty minerals in which the metal is present in more than trace amounts. Mercury-bearing ores have been found on all continents, except Antarctica, and also on the moon. Zones of relatively high mercury content (illustrated in figure 2.1) have been identified on the earth's surface and under the seas; but the distribution is not uniform. Mercury has been estimated to comprise 2.7×10^{-6} per cent of the earth's crust. Data on levels of mercury in various components of the environment are given in Table 2.2.

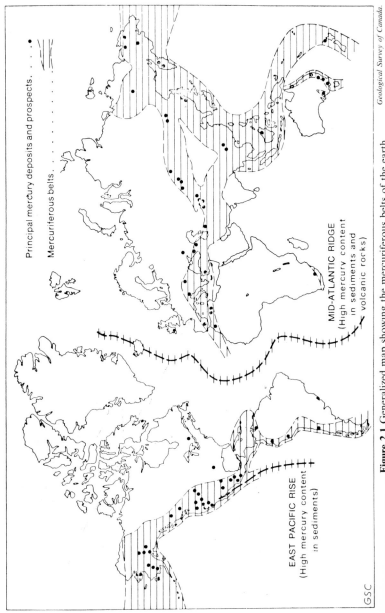

Figure 2.1 Generalized map showing the mercuriferous belts of the earth

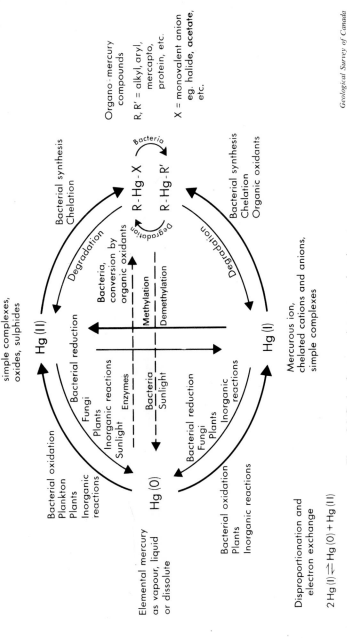

Figure 2.2 Cycle of mercury interconversions in nature

Table 2.2 Some "normal" ranges for total mercury

Lithosphere	0.000 – 10.0 ppm
Hydrosphere	0.01 – 6.0 $\mu g/l$
Atmosphere	0.005 – 1.0 $\mu g/m^3$
Foods (fresh weight)	0.01 – 1.5 ppm
Human urine	1.0 – 25.0 $\mu g/l$
Human blood	1.0 – 50.0 $\mu g/l$
Human hair	1.0 – 5.0 ppm

$\mu g/l$ = parts in 10^9

Sources

Mercury has been mined on a commercial scale in about fifty countries throughout the world, the richest source being at the mines of Almadén in Spain. In America the principal mercury mines are in Canada, Mexico, California and Nevada. Important mines are found in Italy and Yugoslavia. The world production during the period 1950–1970 was somewhere between 200,000 and 300,000 flasks (1 flask = 76 pounds).

Forms and compounds

Mercury forms hundreds of compounds. From a toxicological and environmental point of view the most useful, and commonly accepted, classification is:

Metallic mercury—liquid and vapour
Inorganic salts, such as sulphides, chlorides, nitrates and oxides
Alkyl compounds, such as those containing an ethyl or methyl radical
Alkoxyalkyl compounds, usually of a complex nature
Aryl compounds, particularly the phenylmercurials

Chemistry

Mercury is the only metallic element which is liquid at ordinary temperatures. Its atomic number is 80 and its atomic weight is usually given as 200.59, the latter being based on a mixture of seven naturally-occurring stable isotopes with mass numbers between 196 and 204. It has valences of 1 and 2. As mentioned above, mercury is capable of forming hundreds of compounds, each with its own chemical properties. Even a partial discussion of these properties is beyond the scope of this review but, where the chemistry is related to important environmental concerns or toxicological actions, proper consideration will be given. Mercury readily forms alloys known as amalgams with practically all metals except iron.

Dental fillings are essentially amalgams of mercury and silver. Of some importance is the fact that mercury has a relatively high vapour pressure at ordinary temperatures, the rate of vaporization increasing with increase in temperature.

Mercury release (natural and man-made)

As long ago as the sixteenth century it was suggested that there was a "mercury cycle" whereby the metal circulated throughout the lithosphere (earth's crust), atmosphere, hydrosphere (earth's waters) and biosphere (living organisms—plants and animals). More recently this cycle has been portrayed in various ways such as that shown in figure 2.2.

The question of release and/or translocation is of importance in many respects, but in the present context the greatest importance is related to the potential of released or translocated mercury entering into food chains. As is true of all elements, the total amount of mercury in, on and around the earth is constant. But man and nature can translocate it and convert one form of mercury into another. Man's ingenuity has resulted in the synthesis of hundreds of mercury compounds—far more than have been found to result from natural conversions.

Various estimates have been made of the amounts of mercury released into the environment by man and by nature. It has been calculated, for example, that global weathering of mercury amounts to 800 tonnes per year and that the release of mercury from fossil fuels for the period 1900–1970 was 511×10^3 tonnes. There is a statement to the effect that during the years 1945–1958 in the United States more than 10,000 tonnes of mercury were "lost" to the environment as a result of industrial uses.

Another figure which has been given is that 2.5 kg of mercury per day are discharged into the air by a coal-burning generating plant of 700-MW capacity. Obviously the quantity of mercury released into the air will be in proportion to its concentration in the fuel which is burned and, therefore, can have wide variations. The global figure for mercury release from the burning of coal has been estimated to be 3000 tonnes per year.

Data on man-made releases of mercury into the environment are incomplete and for the most part based on assumptions and projections. It has been claimed, however, that the mercury content of the atmosphere appears to result from the degassing of the earth's crust. Increased flux may come about as a result of the enhancement of this degassing process through the actions of man, such as construction and other activities which disturb the integrity of the crust and thus facilitate the escape (degassing) of mercury vapour. One observer has commented that

... a survey of industrial activity has not revealed mercury releases to the atmosphere that can rival that of the natural degassing rate, estimated to range between 2.5×10^{10} and 1.5×10^{11} g/year.

Volcanic action as a natural source of environmental mercury has been described.

Analytical methods

During the past decade or so, striking advances have been made in a number of aspects of analytical chemistry. Many new techniques have been developed, causing a veritable revolution with replacement of older traditional "wet" methods by new procedures involving delicate expensive instruments and requiring highly skilled technicians. In the environmental field, neutron activation analysis and atomic absorption spectrophotometry have come into prominence in analyzing for total mercury. Gas chromatography and mass spectrometry, particularly the former, are used for identifying and measuring organic mercury, especially methyl mercury compounds. Chemists and physicists are constantly at work refining older methods and devising new ones in an effort to improve sensitivity, specificity and reproducibility.

At the present time the most widely used method for measuring total mercury in environmental samples is flameless atomic absorption spectrophotometry; for organic (methyl) mercury gas-liquid chromatography is commonly employed. Present techniques are capable of measuring mercury down to nanograms and in the parts per thousand million range.*

One of the problems in environmental science has been difficulty in standardizing analytical methods to the point where one laboratory can consistently reproduce the results of another. Improvements in this respect are being made, but most chemists tend to be sceptical of analytical data reported before 1970 or 1971. In analyzing for mercury in biological materials such as blood, urine, hair and foods, it is necessary first to remove extraneous matter to eliminate possible interference. This is difficult to accomplish without loss or addition of mercury and resulting distortion of results. Scrupulous attention must be paid to the purity of reagents and cleanliness of apparatus.

Units of Measurement. It is difficult, even for some trained scientists, to form a meaningful conception of what is meant by a microgram (μg), a nanogram (ng) or a part per million (ppm). We all have a good idea of what a pint of milk or a pound of sugar looks like, but not many will have any idea, for example, of how many micrograms or nanograms are in a pound and what a microgram or nanogram looks like. One pound is approximately 450 grams, or 450 000 milligrams or 450 000 000 micrograms or 450 000 000 000 nanograms. In liquid measure, using the standard apothecary "drop" (called a minim) as the starting point, there are approximately 10 000 drops in a pint, i.e. one drop per pint = one part in 10 000. One part per million then means one drop in about 100 pints = $12\frac{1}{2}$ UK gallons. Tankers used for milk or petrol have capacities up to 5000 or 6000 gallons. Thus one part per thousand million would be less than one drop per tanker.

Toxicological and physiological aspects (including risks to man)

There is confusion in the literature on mercury poisoning because of the failure of some writers to understand the basic principles of toxicology. Most important, perhaps, is the fact that any substance in overwhelming doses can cause serious or even fatal poisoning. This applies to such necessities as oxygen, sodium chloride and water. At the same time, such potent agents as arsenic and strychnine have found important uses in medicine. This is true also of mercury and its compounds. They can kill and they can cure. Arbitrary designation of a substance as "toxic" or "non-toxic" has no scientific basis. The major controlling factors are:

The type of compound (chemical form)
Size of dose
Route of absorption
Duration of exposure
Synergism or antagonism of other agents
Host factors (biochemical individuality)

Toxicologists who work with mammals seem to have been slow in giving adequate recognition to an important concept that has been known to plant physiologists for at least a decade, namely the biphasic nature of the dose-response curve (figure 2.3). This means that very high and very low

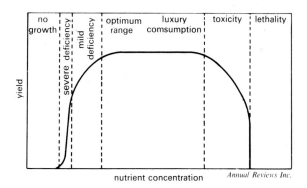

Figure 2.3 Idealized growth of an organism as a function of the concentration of an essential nutrient (after Smith, P. F., 1962. *Ann. Rev. Plant Physiol.* **13**, 81.

concentrations of a substance are incompatible with optimal growth and even with survival, while between the extremes there is a range in which growth will occur. Major applications of this principle are found in the field of micronutrients and at least theoretically in toxicology. They may provide an explanation of why mercury and other heavy metals have a bad reputation.

Recent developments in micro-analysis have made it possible to determine with some degree of accuracy microgram and even nanogram quantities of mercury whereas, formerly, sensitivities were limited to milligrams or major fractions thereof. This meant that mercury levels in biological materials (urine, blood, tissues, plants) could be detected only when they fell on the upper part of the dose-response curve; that is, levels which were harmful or even lethal. Smaller amounts would not be found and hence the analyses would be reported as "zero" or "absent". Detectable amounts would be associated with adverse effects and, therefore, any amount that could be found would be assumed to be dangerous. Emphasis on looking for *bad* effects rather than *any* effect may also have resulted in distorted ideas about mercury. The possibility that mercury may have essential functions in living systems is slowly gaining acceptance.

Specific toxicological effects
Metallic mercury
In its liquid form, metallic mercury can enter the body by swallowing, through the skin, or by injection. For many years oral doses of quicksilver were used in treating a variety of diseases: repeated administration of a few grams for everything from apoplexy to worms, and amounts up to a kilogram in a single dose for intestinal obstruction. Starting around the year 1500 and for about 450 years thereafter, mercury mixed with a fatty base and rubbed into the skin was one of the most important measures in the treatment of syphilis. There is conflicting evidence as to whether this did more harm than good.

Metallic vapour
Because of its volatility at ordinary temperatures, spilled mercury always constitutes a potential hazard. Exposures to low levels of vapour over periods of months or years can result in chronic poisoning. The classical symptoms in such cases are gingivitis (inflamed gums), tremors and personality disturbances (erethism). Kidney injury, manifested by proteinuria (albumin in urine), is occasionally seen, and there is almost invariably an elevation of mercury levels in blood and urine above the "normal" concentrations of 50 and 25 $\mu g/l$ respectively. Complete recovery following termination of exposure is normal.

Inorganic salts
Contributing to mercury's bad reputation is the erstwhile use of corrosive

sublimate (bichloride of mercury) by suicides. A gram or two taken by mouth will cause severe vomiting and diarrhoea, bleeding from the intestinal tract, profound injury to the kidneys, possibly kidney failure, and death by uremia. Cases of this type are now rarely seen. Mercurous chloride (calomel) is less soluble than the mercuric salt and hence less harmful if swallowed. Ammoniated mercuric chloride and mercuric oxides have been used in ointments. Other inorganic compounds are of relatively minor importance.

Organic compounds—general comments

Articles still appear in which authors treat all organomercurials as a single class. From a toxicological point of view, nothing could be more misleading; organic compounds of mercury span the entire spectrum from life-saving therapeutic agents to highly lethal chemicals. The importance of specificity in any detailed discussion of mercury toxicity should be emphasized. Organomercurials which are useful as diuretics in treating water retention (dropsy) have been found to be unstable, readily liberating mercury ions (Hg^{2+}). Aryl (phenyl) compounds behave in a similar manner. Alkylmercurials tend to be more stable. These differences in stability can explain at least some of the differences in toxicity among the organomercurials.

Arylmercurials. This class of compound embraces all mercury derivatives of aromatic hydrocarbons including such groups as benzyls, tolyls, naphthyls, and many others, as well as their halogenated and other forms. Many hundreds of varieties of arylmercurials have been synthesized, but only those containing a phenyl group are of major importance, especially as pesticides and biocides. Phenylmercuric acetate (PMA) is the most widely used of the arylmercurials, but the benzoate, lactate, nitrate, oleate, propionate and others have found a place in the pesticide field. There is fairly good evidence that for humans the toxicological behaviour of the various phenylmercurials is similar.

Extensive studies of human exposure to phenylmercurials conducted at Columbia University led to the following conclusions:

1. For human exposure there appears to be no significant difference in the toxicity of the various phenylmercurials.
2. Phenylmercurials may be absorbed through the intact skin or mucous membranes, but relatively high concentrations must be applied before measurable amounts will be absorbed in this way.
3. Solutions of phenylmercurials in concentrations in the per cent range will regularly cause second-degree chemical burns of the skin (blistering), but more dilute solutions are only slightly irritating and skin sensitization rarely occurs.

4. Persons may be occupationally exposed to phenylmercurials in concentrations many times the accepted Threshold Limit Value of 0.05 mg/m³ in air for many years without showing evidence of poisoning, even though they absorb sufficient of the compounds to cause them to excrete mercury in their urine in the milligram per litre range.
5 Poisoning due to phenylmercurials is extremely rare, and chronic occupational poisoning is unknown.

Alkoxyalkylmercurials. Compared with the aryl and alkyl compounds of mercury, relatively little is known about the toxicity of the alkoxys. Those which are related chemically to the mercurial diuretics are, of course, practically non-toxic. Limited experimental work has shown that methoxyethylmercuric chloride resembles PMA in its behaviour, and that it is distinctly less toxic than the alkyl equivalent.

Alkylmercurials. A great number of articles on alkylmercurials have been published since 1970. Episodes of poisoning at Minamata and Niigata in Japan, and in Iraq, Pakistan, Guatemala and the United States have been widely publicized, particularly Minamata disease resulting from the consumption of contaminated fish. First in Japan and Sweden, and later in North America, extensive studies of the toxicology of methyl mercury and other alkyl compounds have been conducted leading to the following conclusions:

1. Alkyls are distinctly more toxic than aryls, alkoxys and inorganic compounds of mercury.
2. Fatal alkylmercury occupational poisoning through inhalation and non-occupational poisoning through ingestion have occurred.
3. Alkylmercurials are readily absorbed from the intestinal tract and through skin. They can pass through the placenta and cause foetal injury.
4. The principal site of injury from alkyls is the central nervous system. Damage, when it occurs, tends to be permanent. The most frequent disturbances are in vision, hearing, sensation, muscle function and mentality.
5. Alkylmercurials show affinity for lipid tissues and when taken up are liberated slowly. The "half-time" of retention in man is about 70 days for the total human body. Calculation of this retention time averages out the longer period for the brain and shorter times for other tissues.
6. There is no known effective treatment for alkylmercury poisoning. Prolonged physical therapy may result in some restitution of function.

Methyl mercury in fish

Because of the widespread concern over possible harm to humans who eat species of fish, it seems desirable to discuss this subject in some detail.

There can be little doubt that methyl mercury played a significant role in the Minamata tragedy, yet it has been claimed by at least one investigator that something in the fish other than the mercury was the responsible agent.

It has been shown that the water of Minamata Bay was polluted by a number of chemicals in addition to organic and inorganic mercury, specifically arsenic, thallium, manganese, selenium and aldehydes.

Biologically methylated mercury may not have the same toxicological properties as the chemically synthesized compound. There is evidence, both theoretical and experimental, to support this conclusion. Marine micro-organisms have been shown to be able to detoxify mercury compounds under some conditions.

On the other hand, feeding experiments using cats and freshwater fish have shown that "natural" and synthetic methyl mercury may behave in a similar manner in producing toxic effects. At the time of writing there is still uncertainty about whether or not this is true in man. Also uncertain is whether marine and freshwater fish are similar in this respect. An important factor may be the amounts of selenium present, there generally being higher levels in saltwater fish.

Widespread publicity has been given to a woman in New York who allegedly suffered from methyl mercury poisoning as a result of eating large amounts of swordfish. Less attention has been paid to the doubts expressed by the physician who reported the case, and no publicity has been given to his most recent views that ". . .conclusive documentation of the cause of Mrs. N.Y.'s symptomatology will always elude us . . ." and that ". . .there were a number of bits of information which could have solidified the diagnosis but were not present and are still not present." Certainly there is room for doubt in this case.

As mentioned elsewhere, mercury is ubiquitous and hence may be considered to be normally present in fish as well as in all articles of diet. Two species, swordfish and tuna, have commanded particular attention—the former because of the relatively high mercury levels, and the latter because of its popularity as a food and hence the large amount which is consumed by humans (and cats). It has been shown repeatedly that concentrations of mercury increase with progression from lower to higher levels in the food chain, and that those in fish are directly related to the age and size of the fish. This offers at least partial explanation of the findings in swordfish: those caught for sale as food usually weigh several hundred pounds. Since very few people are likely to eat swordfish as often as once a month, it borders on the absurd to be concerned about mercury poisoning from this source.

Analyses of fish as much as 90 years old (museum specimens) have shown that there appears to be no difference in the mercury concentrations in these ancient fish and those being found today, thus indicating no recent build-up of mercury as a result of man-made contamination.

The status of methyl mercury

From the standpoint of human health, the role of methyl mercury is of overriding importance and concern. This is understandable in the light of the tragic episodes of Minamata, Niigata, Iraq and elsewhere. Much has been learned and is still being learned from studies of these mass poisonings with methyl mercury compounds. New reports continue to become available, necessitating continuing reappraisal of accepted beliefs. Some of the points which require this continuing reappraisal are:

1. *Analytical Difficulties.* Chemists who have undertaken the micro-analysis of environmental and biological samples for mercury are practically unanimous in emphasizing the difficulties in obtaining results which are reproducible either in their own or in other laboratories. There is general agreement that many of the data reported prior to about 1970 are suspect, and even some of the more recent reports are open to question. Noteworthy are the differences in results of analyses for methyl mercury in fish reported from Sweden, Japan and the United States. Not without reason is caution exercised by knowledgeable research workers in the interpretation and use of analytical data.

   Even though some chemists claim that their analytical results are reproducible, these seem to be the exception rather than the rule, particularly in regard to methyl mercury. This position is supported by several published reports of "round robin" studies in which the same specimens were analyzed in several laboratories. The situation is improving, but still not entirely satisfactory. Older data must remain suspect.

2. *Methylation of Mercury at Minamata.* Because of misinformation given out by a local chemical plant at Minamata, it was originally believed that the cause of poisoning of local residents was biomethylation of inorganic mercury. Later it was learned that the plant was discharging the chemical methyl mercury in its effluent. When this became known, the Japanese scientists who were investigating the outbreak rejected their original theory and attributed the poisonings to this direct discharge of methyl mercury.

3. *Human Poisoning by Methyl Mercury.* As was the case at Minamata, all authenticated cases of methyl mercury poisoning have been a result of the use of foods which have been grossly contaminated extrinsically with a methyl mercury compound. There have been no cases in which the methyl mercury was present as a result of biological processes. With the exceptions of Minamata and Niigata, studies of human populations in which there has been large-scale ingestion of fish taken from waters containing "excessive" amounts of mercury have not shown any evidence of the slightest adverse effects.

The most comprehensive of the population studies are those conducted by a research team from the University of Rochester, New York, on Peruvian fishermen and on Samoan cannery workers and fishermen. All of these groups consistently consume fish as a major item of diet, and the mercury content of the diet is sufficiently high to produce significant elevations of mercury in blood and hair. In spite of this, no evidence of mercury effects was observed. These findings, in addition to those of earlier studies, support the conclusion that mercury found in fish caught in their natural habitats is of no toxicological significance. Only where there has been gross chemical contamination of water, as at Minamata, is there a real danger.

In the Minamata and Niigata episodes the dire effects were due in part, as mentioned above, to the contamination of fish by chemical methyl mercury which had been discharged into bodies of water. In addition, there is reason to believe that the nature of the diets of the victims was of importance. Not only did they consume contaminated fish as their principal source of protein food, but this one-sided diet possibly resulted in deficiencies in other respects, thus making the population more susceptible to poisoning.

Studies of the mercury levels in water and fish under pristine conditions have shown that even when man-made contamination is virtually impossible, levels may exceed those which some people consider safe or permissible.

Methylation and Demethylation of Mercury. Some of the most important research of the past two or three years has been devoted to studies of methylation and demethylation as it takes place in the natural environment. Originally it was believed that methylation of mercury was a one-way reaction which could affect all forms of mercury. This belief is no longer tenable. It has now been demonstrated that methylation can occur in nature only under very specific and limited conditions of mercury concentration, acidity, temperature, bacterial species, and that not all mercurials are converted to the methyl forms. Phenylmercuric acetate, for example, is converted to diphenyl mercury. Even more important are findings published in 1972 and later showing that there are demethylating processes present in the environment under a wide range of conditions, and that these are constantly at work reversing the methylation and thus preventing excessive or dangerous accumulations of methyl mercury (see figure 2.2).

Nature's Protective Mechanisms. In addition to demethylation, nature provides a variety of other protective mechanisms. Alterations of

mercury toxicity through bindings in protein molecules in animals ha
been mentioned. Similar bindings have been demonstrated in th
organic matter naturally and normally present in soils and silts, th
humates and fulvates in particular. It has been shown that algae unde
some conditions can detoxify mercury compounds. The elemen
selenium. normally present in all living organisms, has been found t
exert an antidotal effect on ingested mercury. Manganese, which i
present in all waters, has been shown to be an avid scavenger of mercury
and iron has a similar action.

Studies on trace-metal metabolism have shown that there ar
innumerable interrelationships among those for which essentia
functions have been found, and that similar relationships probably exis
for all. Under conditions as they exist in nature, combinations of al
kinds are present, and there is no doubt that any effects (or lack o
effects) result from very complicated interrelationships. Gros
dislocations of natural balances can be dangerous, but they are likely t
be found only in rare instances, such as in the Minamata episode. If th
protective balance provided in nature did not exist, all or most life o
earth would be impossible. Extreme distortions of those balances whic
occur in nature, whether through addition or subtraction of element
normally present, may be equally dangerous.

6. *Development of Tolerance.* All living things have a capacity fo
developing tolerances toward external stresses, including chemica
stress. These so-called adaptations are effective in providing protectio
against foreign chemicals when exposure is in small increments, but the
break down under massive attack. Evolutionary evidence, strengthene
by clinical and experimental observations, supports the theory that lif
has survived on earth by virtue of being able to adapt to all element
present in the primordial environment. Gradual increments cause n
harm and, over a period of time, tolerance changes to dependence
Research directed to testing this hypothesis has led to the discovery tha
a number of elements, including chromium, manganese, selenium an
others, once thought to be non-essential or even harmful, actually hav
vital functions in man, other animals and plants.

7. *Genetic Damage.* One of the most frightening of the alleged effects o
mercury, especially methyl mercury, is the possibility of geneti
(inheritable) injury. A significant comment on this was published in 197
by scientists at Kumamoto University in Japan as part of their follow-u
on cases of Minamata disease. They state that "...in our research surve
in Minamata district at this time we could not find any evidence o

genetic damage or genetic variation." Chromosomal abnormalities were no more frequent in the exposed than in the control population. This negates the original allegation of genetic damage which had been made on the basis of experimental studies on fruit flies and onion tips. Birth defects due to injury occurring *after conception* are found in the offspring of severely poisoned mothers.

Hydrosphere-Biosphere

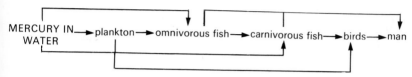

Pedosphere-Biosphere

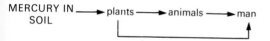

Atmosphere-Biosphere

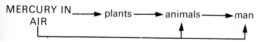

Figure 2.4 Generalized food chains according to the Geological Survey of Canada.

Human intake of mercury from food and other sources

The ubiquity of mercury, mentioned above, means that all humans will absorb a certain amount of the element from food, water and air. Of these three sources only food is of any consequence under ordinary circumstances. Absorption through the air can be significant in occupational exposures and, theoretically, in cases of gross contamination. Water could be important, but for practical purposes the last two of these sources need not be considered (see figure 2.4).

The first systematic study of mercury in food was done by the German chemist Stock around 1934. At that time it was estimated that the average daily intake was about 5.0 μg.

F

In recent years two agencies of the United Nations, the Food and Agricultural Organization (FAO) and the World Health Organization (WHO) have devoted considerable attention to mercury in foods and have published a number of reports. These usually have been the result of joint FAO/WHO action. As early as 1963 the FAO/WHO Codex Alimentarius Commission recommended an upper level of 0.05 mg/kg of mercury for all foods except fish and shellfish. This figure is equivalent to 0.05 ppm or 50 ng/g.

Experts convened by FAO/WHO in November 1972 considered the use of mercurial pesticides in agriculture as possible sources of mercury in foods. They noted, among other things, that

the uptake of mercury into crops from dressed seed was insignificant as a potential source of food contamination

but recommended that

Further study should be given to the question of replacing them, wherever possible, by compounds less likely to produce such poisoning incidents

as those which occurred in Iraq in 1972 when seed grain was diverted to consumption by man, poisoning thousands and killing hundreds. This group of experts reaffirmed the views of another FAO/WHO Committee which in 1967–1968 had suggested tolerance limits for organomercurials in foods of 0.02–0.05 ppm.

A number of scientists, notably those in Sweden, have attempted to establish an "acceptable daily intake" (ADI) for mercury in foodstuffs, particularly in fish. Unfortunately there was a scarcity of scientifically valid data that might serve as a basis for setting the ADI. This necessitated the use of a number of assumptions and extrapolations. The Swedish scientists, for example, in commenting on the data used by the US Food and Drug Administration (FDA) in setting the guidlines of 0.5 ppm of mercury in fish, state that "There are several uncertainties in the estimation. . ." of the extrapolations which led to the 0.5 ppm figure and they refer to the basic data as ". . .scanty and unreliable."

A report published by the Massachusetts Department of Public Health in 1972 includes a discussion of the Swedish attempts to set an ADI for mercury in the diet. The authors accept the fact that all mercury in fish is in methyl form and that this is the most toxic of all mercury compounds. They point out that a person would have to consume at least three pounds of fish containing 0.5 ppm of mercury every week in order to exceed the ADI set by USFDA.

Since 1971 the USFDA has been studying the mercury content of the "average" American diet. Their calculations show that in 1971 the average daily intake of mercury was 0.00248 mg/day and for 1972 it was 0.00385 mg/day. Expressed as micrograms these figures are 2.48 and 3.85

respectively, levels which are close to, but lower than, the 5.0 μg/day estimated by Stock in 1934 in Germany. Swedish experts have stated that up to 100 μg (0.1 mg) is a safe ADI for mercury in food.

Environmental standards for mercury

Standards may be defined in a number of different ways. As applied to the environment the term "standard" designates any definite rule, principle or measure established by authority. Of major concern is the necessity to employ quantitative factors (numbers) in promulgating environmental standards; without these numbers, bases for enforcement would be difficult or impossible to establish. The fact that standards are set by "authority," whether this be official, quasi-official or voluntary, does not necessarily mean that they are fair, equitable, or based on sound scientific knowledge. They may have been, and all too often are, established arbitrarily on the basis of inadequate technical data, perhaps tempered by a cautious factor of safety.

Unfortunately a standard may be regarded by some persons as an end in itself, a mere number to be met, rather than as a means to be used toward achieving an appropriate degree of environmental quality. This is not to say that standards are not important, in fact necessary, to protect the health of various segments of the population. Their validity, and hence their efficacy, depend to a considerable degree on an understanding of their weakness as well as their strength.

Bases for standards. By their very nature, environmental standards almost invariably include specific numerical values which are used as a dividing line between what is acceptable and what is, not. Presumably someone will benefit in some way from the imposition of standards; at the same time, someone will have to bear the costs of complying with the standards. Some of the anticipated benefits may be nebulous; the costs are not. Possible threat to human health is perhaps the most cogent, but not the only, reason for imposing environmental standards. Other animals, vegetation, inanimate objects and aesthetics must be considered.

Various bases, alone or in combination, may be or have been used in establishing environmental standards for mercury. These include:

1. *Established Practice.* This may be applicable where existing levels are considered acceptable and where it is not necessary or desirable to undertake extensive changes in existing control measures.
2. *Attainability.* This presupposes the existence of suitable technology for both control and measurement, as well as economic feasibility and enforceability.
3. *Experimentation.* In the absence of existing data it may be necessary to set up controlled experiments using laboratory animals, plants, ecological communities or human volunteers.

4. *Epidemiological Studies.* These may involve the study of "exposed" populations over periods of months or years, or of the effects of acute exposures resulting from natural or man-made catastrophes.
5. *Mathematical Models.* These may use statistical or computerized techniques.
6. *Educated Guesses.* Presumably these would make use of all available information, but they cannot ordinarily avoid a large subjective element.
7. *Analogy and Extrapolation.* These are temptingly attractive but are fraught with danger, frequently leading to what is little more than speculation.

Many of the basic principles outlined above will find application in discussions of specific standards.

Just as in medicine it is not possible to know what is abnormal without knowing what is normal, so in the setting of standards there must be an analogous point of reference. In dealing with the environment, concepts of "normal" are by no means simple. Refinements in analytical techniques have led to the realization that mercury (as well as many other elements) is ubiquitous; at least it has been found in every part of the environment in which it has been sought. This means that there are certain background levels which are present as a result of natural processes, regardless of man-made modifications. These background levels vary from place to place, and probably from time to time in the same places. This may perhaps properly be termed "natural." If a background level should be altered as the result of a phenomenon such as drought, flood or volcanic eruption, the result would still be natural (i.e. not caused by man) but might not properly be considered "normal."

Under pristine conditions all living things, including man, will absorb some mercury from food, water, and air. Here, too, the "natural" amounts can vary, depending on what is present in nature. Many living organisms, from algae to mammals, tend naturally to concentrate the mercury they receive from their environment. This would properly be accepted as "natural", but there is some question as to the extent to which this concentration can go and still be called "normal." A dividing line might be drawn at the point where absorption results in some physiological disturbance. This interpretation would be in accordance with the recognized existence of naturally occuring toxins in plants and animals. In any event it is important to realize that "natural" and "normal" are not necessarily the same, that there is no fixed point for either "natural" or "normal" mercury, and that abnormal is not necessarily synonymous with pathological or dangerous.

Current and anticipated standards call for the measurement of mercury in the parts per thousand million or nanogram range. This type of microanalysis is not something that can be performed routinely in just any laboratory. Furthermore, as mentioned above, it cannot be assumed that all analytical methods will yield comparable results, nor that different laboratories using the same method will obtain the same findings on aliquots of the same sample. It follows, therefore, that the development and enforcement of any standard for mercury in the environment must include provisions for standardization of analytical techniques and continuing monitoring of laboratories for reliability of results.

"Normal" values for mercury. Perhaps the most important of all standards are those which are related to concepts of what is "normal." Sometimes the terms "background level" or "natural" are used, applying to concentrations of mercury in the general atmosphere, hydrosphere, lithosphere and biosphere due to "natural" processes without man-made intervention. Mercury, although ubiquitous, is by no means uniform in its distribution. The presence of mercury-rich ores obviously can produce elevated natural background levels; therefore, what is "natural" or "normal" must be evaluated with consideration being given to geologic formations. Some "normal" ranges for mercury are given in Table 2.2. The values listed apply only to those countries or regions in which tests have been performed, and do not necessarily represent what might be "normal" elsewhere.

Water
Drinking water. Concern about mercury in drinking water is of recent origin, there being no mention of this metal in the widely-used US Public Health Service Drinking Water Standards of 1962. In fact, as late as 1969, an extensive survey of drinking water supplies in the United States did not include mercury among the substances that were evaluated. As far as can be determined, the Soviet Union was the first nation to set a standard for mercury in drinking water, this having been done around 1950. The standard is 0.005 mg/l for mercury or mercuric ions which is equivalent to 5.0 parts per thousand million. This same figure was unofficially adopted in the United States but subsequently a level of 0.002 mg/l was proposed. Public apprehension over the quality of public water supplies in the United States, stimulated to a great extent by irresponsible news media, has led to a phenomenal growth in sales of bottled waters which allegedly come from "pure" springs. Responding to this development, the US Government in March 1973 issued proposed standards for bottled water, but did not include mercury among the chemicals to be considered, an oversight which was not corrected when final standards were adopted.

The third edition of the *International Standards for Drinking Water* (World Health Organization, 1971) recommended a tentative upper limit for mercury in drinking water of 1.0 μg/l. It was noted that this figure is related to levels found in natural waters used for drinking purposes. This seems to be a sensible realistic standard, preferable to that of the Soviet Union and that being proposed for the United States. The WHO standard is based on an assumed consumption of 2.5 litres of water per person per day, a figure which may be higher than that which is usual in some countries.

Waste water. Since many communities derive their drinking water from streams which have received the sewage (waste water) of other communities, it is a matter of general concern to control the quality of fluid wastes. This question takes on added importance in view of the paucity of information on the extent to which mercury can be added or subtracted during the ordinary processes of water purification designed to produce safe drinking water.

Japan was among the first nations to establish an effluent standard when in 1969 the government set a limit of 10 $\mu g/l$ for methyl mercury in waste water. At least one political unit of the United States, the State of Illinois, has established a waste water standard to "eliminate all measurable, non-natural concentrations of mercury from whatever source." At public hearings held before the standard was adopted, it had been pointed out that the provisional drinking water standard allowed ten times as much mercury as was to be permitted in sewage. In another case involving similar restrictions a member of the U.S. Supreme Court commented that ". . .this meant that the only thing that could be discharged into water was pure water."

Air

Industrial air. Standards for permissible concentrations of atmospheric contaminants in factories, mines and other places of work have been known as Maximum Allowable Concentrations (MAC) and more recently as Threshold Limit Values (TLV). Attempts to develop such standards began in the United States in the 1920s, but for mercury the first definite proposals came during the following decade. As a result of extensive studies made by the US Public Health Service in the hatters' fur-cutting industry and the felt-hat industry, a recommendation was put forth that exposure to atmospheric mercury should not exceed ". . .1.0 mg Hg per 10 m^3 of air. . . (which) probably represents the upper limit of safe exposure." (This value, of course, is the same as 0.1 mg per m^3). The validity of this standard has been questioned largely because, in the USPHS reports, cases of intoxication due to mercury were described among workers exposed at this level.

For a number of years a TLV of 0.1 mg per m^3 for metallic mercury vapour and other inorganic mercurials was accepted in the United States and elsewhere, but lower values are now current as shown in Table 2.3

Table 2.3 Some threshold limit values for occupational exposure to mercury (compiled from various sources 1969-1973. Values in mg/m³ of air.)

Country	Inorganic	Alkyl
Czechoslovakia	0.05	
Germany		
Democratic Republic	0.1	
Federal Republic	0.1	
Great Britain	0.1	
Hungary	0.02	
Japan	0.05	
Poland	0.01	
USA	0.05*	0.01
USSR	0.01**	0.005**

Includes aryl compounds
*Ceiling values

scientists meeting in Stockholm in 1968 made the following recommendations:

- Methyl and ethyl mercury salts. No air concentration is recommended. The mercury level in whole blood should not exceed 10 μg Hg/100 ml (as total Hg). This blood concentration is a ceiling value, and it should not usually be exceeded with continuous 8-hour exposure to 0.01 mg/m³ of alkyl mercury in air.
- Mercury vapour: 0.05 mg/m³
- Inorganic mercury salts and phenyl amd methoxyethyl salts: 0.1 mg Hg/m³.

Ambient air. To the uninitiated, some of the terminology employed in the field of air pollution can be confusing. The term *ambient air* refers to the general atmosphere as distinguished from the air in confined spaces such as in factories. Ambient air standards, therefore, deal with quantitative and qualitative factors in the air breathed by the general population at all times (Table 2.4). Air quality criteria refer to the amounts of pollution and the duration of periods of exposure (concentration × time) at which specific adverse effects on health and comfort may occur. An air quality standard establishes the level of a pollutant in the ambient air that cannot be legally exceeded during a specified time in any given geographical area. Attempts to meet air quality standards are made through the imposition of emission standards which prescribe the maximum amount of a pollutant which can legally be discharged from a single source (Table 2.5).

Food

The presence of mercury in food has been recognized for more than forty years, but it was not until the 1960s that any serious attention was paid to establishing standards for acceptable concentrations. At the first session of

Table 2.4   Some ambient air quality standards for mercury as Hg (24-hour averages); from Martin, W. and Stern, A. C. Data presented at Inter-Regional Symposium on the Use of Air Quality Criteria in National Air Pollution Control Programmes, WHO, Geneva, December 10–14, 1973.

Country	Standard (μg/m^3)
Bulgaria	0.3
Germany (Democratic Republic)	0.3
Israel (tentative)	10
Romania	10
Soviet Union	0.3

Table 2.5   Some emission standards from stationary sources for mercury; from Martin, W. and Stern, A. C. Data presented at Inter-Regional Symposium on the Use of Air Quality Criteria in National Air Pollution Control Programmes, WHO, Geneva, December 10–14, 1973.

Country	Source	Standard
Australia		
New South Wales	Any	0.2 g/m^3
Queensland	Any	0.01 grains/ft^3
Victoria	Any	0.01 grains/ft^3
Czechoslovakia	Metallic	0.003 kg/h
Singapore	Any	0.02 g/m^3
Sweden	Chlorine mfr.	0.001 kg/ton Cl
	Vented hydrogen	0.0005 kg/ton Cl
USA	Chlor-alkali plants	2.3 kg/day
	Ore processing	2.3 kg/day

the joint FAO/WHO Codex Alimentarius Commission in Rome 1963 tolerable levels for mercury in certain foods were set at 0.05 part per million for total mercury. This standard specifically excluded drinking water, fish and shellfish, for which no standard was proposed. In 1968 the Commission took up the question of organo-mercurial compounds in foods and concluded at that time that

It is not possible to establish an acceptable daily intake on the basis of available information. Any use of mercury compounds that increases the level of mercury in food should be strongly discouraged.

A similar objective had been espoused by the US Food and Drug Administration in 1938 when it set a "zero" tolerance for residues of mercurial pesticides on foods. (The meaning of "zero" is not defined in the FDA regulations).

Table 2.6 Some standards for mercury in foods

Country	Type of Food	Tolerance
Australia		
Victoria	All (?)	0.1 ppm
S. Australia	All (?)	0.1 ppm
W. Australia	All (?)	0.01 ppm
Benelux	All (?)	0.03 ppm
Brazil	All (?)	0.05 ppm
Canada	Fish	0.5 ppm
Denmark	All	0.05 ppm
Germany	Fruits, vegetables	"zero"
Japan	Fish	1.0 ppm*
New Zealand	Fruits, vegetables	0.05
Sweden	Fish	1.0 ppm
	Other foods	0.05 ppm
USA	Fruits, vegetables	"zero"
	Fish	0.5 ppm
WHO	All	0.3 mg/wk total
		0.2 mg/wk methyl

*Not an absolute standard.

The "Great Mercury Scare" of the early 1970s resulted in considerable attention being directed toward the question of acceptable concentrations (standards) for mercury in foodstuffs, particularly in fish. As early as 1966 the Swedish National Institute of Public Health had suggested 0.5 ppm (wet weight) as the upper safe level for mercury in fish, this figure being predicated on the assumption that all, or practically all, mercury in fish is in a methylated form. Within a few months it was found that lakes which were a major source of supply were yielding fish with mercury content in excess of the 0.5 ppm, so in February 1967 the permissible level was raised to 1.0 ppm. This is an example of a standard being set on economic rather than scientific grounds. In raising the "safe" level, the Swedish authorities recommended that fish taken from certain lakes be not eaten more than once a week.

Early in 1970, the Food and Drug Directorate of the Canadian Department of National Health and Welfare adopted a level of 0.5 ppm (wet weight) as the upper limit for mercury in fish which was to be offered for sale. This restriction did not apply to fish caught by an individual for his own consumption nor to other foods. According to the Deputy Director-General of the Food and Drug Directorate, the standard was chosen after a review of short- and long-term toxicity studies, consideration of the fish-eating habits of Canadians, available evidence from human investigations in Japan and Sweden, and the example set by Sweden. The Canadian standard admittedly was made on the basis of ". . .data which are deficient in many respects. . ." and as a temporary measure. In discussing the standard at a meeting held in Toronto in Febraury 1971, a Canadian official stated that

It must be emphasized that the present guideline of 0.5 ppm for fish is just that—a guideline. It is not immutable; it is susceptible to change as scientific information about toxicity of mercury becomes more complete. For the moment, it's the best we can do.

Canada's standard is one of expediency.

Due in large measure to the intemperate sensationalistic handling of the mercury story by news media in the United States, government officials came under great pressure to establish some sort of standard for mercury in fish. *Faute de mieux,* they turned to Canada and adopted their standard of 0.5 ppm as a "guideline". (This euphemism offered little comfort to the tuna canning industry, which was obliged to recall millions of cans of their product or to the swordfishermen who found themselves abruptly deprived of a livelihood.) The exact date of adoption of the US standard is not clear, but it probably went into effect in April 1970. Its selection was based on convenience (borrowing from Canada) and on political considerations. Legitimacy for the standard was established by a commission which visited Sweden and Finland for ten days in August 1970.

Due to the tragedies of Minamata and Niigata, standards for mercury in food, particularly in fish, became a matter of major concern in Japan in the late 1960s. Extensive studies of the mercury content of fish were initiated with special attention to that which was present in methylated form. Findings in several Japanese laboratories did not confirm those in Sweden, in that the proportions of methyl to total mercury were found to be as little as 4 per cent, and the average in thousands of determinations to be about 50 per cent. Influenced by these findings, Japanese health authorities placed restrictions on the sale of fish if at least 20 per cent of a minimum sample of 25 fish from a given catch showed more than 1.0 ppm of mercury with analyses being done by more than two laboratories which showed consistent results.

The Japanese approach to setting of standards has much to commend it, particularly the requirement for analyses of a number of fish and the emphasis on consistent laboratory findings. Standardization of analytical nethods and consistency of analytical results must be included as an essential part of any acceptable standard for mercury in fish or other foods.

Following a meeting held in Geneva in April 1972, the Joint FAO/WHO Expert Committee of Food Additives issued its Sixteenth Report in which it is stated:

The Committee established a *provisional tolerable weekly intake* of 0.3 mg of total mercury per person, of which no more than 0.2 mg should be present as methyl mercury, CH_3Hg^+ (expressed as mercury); these amounts are equivalent to 0.005 mg and 0.0033 mg, respectively, per kg body weight. Where the total mercury intake in the diet is found to exceed 0.3 mg per week, the level of methyl mercury compounds should also be investigated. If the excessive intake is attributable entirely to inorganic mercury, the above provisional limit for total mercury no longer applies and will need to be reassessed in the light of all prevailing circumstances.

Conclusions

Mercury is unique in that it is alone among the metals in being a liquid at ordinary temperatures. As one of the "tria prima" of the ancient alchemists, it has long been associated with magic and mystery. Otherwise, mercury possesses most of the characteristics common to all metals: it exists in nature in many different forms; chemists have manipulated it into hundreds of new compounds; in some forms and in some doses it runs the gamut from being, as Woodall said in 1639, "a wicked murderer" to being "a medicine". The irrational fears surrounding mercury in the early 1970s may well be an atavistic phenomenon, harking back to the days of the alchemists. If, as Cheyne advised, mercury is "judiciously manag'd," it need be no more frightening than many other chemicals in everyday use. Sufficient information is available to serve as a guide in achieving judicious management. Mankind would lose much if mercury were to be unfairly condemned and its uses unnecessarily restricted.

FURTHER READING

Friberg, L., and Vostal, J., eds. (1971), *Mercury in the Environment—A Toxicological and Epidemiological Appraisal,* Stockholm, Sweden: The Karolinska Institute (in English).

Gavis, J., and Ferguson, J. F. (1972), Review paper: "The Cycling of Mercury through the Environment," *Water Research* **6:** 989–1008.

Goldwater, L. J. (1972), *Mercury: A History of Quicksilver,* Baltimore, Maryland, USA, York Press.

Hartung, R., and Dinman, B. D., eds., (1972), *Environmental Mercury Contamination.* Ann Arbor, Michigan, USA: Ann Arbor Science Publishers.

Krenkel, P. A. (1973), "Mercury: Environmental Considerations, Part I, "in *Critical Reviews in Environmental Control,* Vol. 3, Issue 3, Cleveland, Ohio, USA, CRC Press.

CHAPTER THREE

LEAD

M. R. MOORE, B. C. CAMPBELL AND A. GOLDBERG.

Introduction

Lead is a grey, ductile, malleable metal, the physical and chemical properties of which ensure that, providing supplies are not exhausted, its future technology will match its history. It is also probable that because of its proximity to man there will be continuing problems relating to its toxicity.

Lead exists only in relatively small quantities in the earth's crust (about $10^{-3}\%$ by weight), yet it has become relatively well known because of the ease with which it may be obtained from its ores, and consequently because of its technical importance to society. Its physical properties are listed in Table 3.1. In natural abundance, lead is a mixture of the four stable isotopes of atomic weights 204, 206, 207 and 208, the ratios of which are dependent on its geological origin. There are in addition many radioactive isotopes found both in natural radioactive decay series and produced artificially in nuclear reactors or charged particle accelerators. Of the radioactive isotopes, ^{203}Pb and ^{210}Pb have been most used in biological trace studies (Table 3.2).

Lead is obtained primarily from its ore galena (lead sulphide) (figure 3.1 and Table 3.3); other ores such as cerussite (carbonate), anglesite (sulphate) and plumbojarosite (hexa-ferric hydroxy sulphate) are weathered products of galena. These ores are mined principally in the United States and the Soviet Union, although mining in Australia, Canada, Mexico, Peru and many other parts of the world accounts for a considerable output. These resources are being mined at a rate of 3.5×10^9 kg per annum (1974). The ores are purified after mining by crushing and concentration, either by water gravitation or froth flotation.

Table 3.1 The properties of lead

Atomic number	82
Atomic arrangement	Face-centred cubic
Interatomic distance	0·349 nm (3·49Å)
Atomic weight	207·2 (lead derived from radioactive sources may have atomic weights varying from this figure)
Density at 20°C	11343 kg m^{-3} 11·34 g/cm^3) (708 lb/ft^3)
Melting point	327·502°C (600·66 K)
Latent heat of fusion	26·2 kJ/kg (6·26 cal/g) (11·27 B.Th.U/lb)
Thermal conductivity 0°C	36 W/(m K) (0·083 cal/(cm s K)
100°C	34 W/(m K) (0·081 cal/(cm s K)
Specific heat (0°C to 100°C)	(127 J/(kg K) 0·031 cal/g/°C(average)
Coefficient of linear expansion	0·000029/°C

Table 3.2 The isotopes of lead. (Since the stable isotopes 206, 207 and 208 are obtained from radioactive decay series the ratios of these can vary according to the geological source and age.)

Nuclide	Half life	Decay mode
^{194}Pb	11 min	γ
^{195}Pb	17 min	EC
^{196}Pb	37 min	EC, γ
^{197}Pb	42 min	EC
^{197}Pbm	42 min	
^{198}Pb	2·4h	EC, γ
^{198}Pbm	25 min	
^{199}Pb	90 min	EC, β, γ
^{199}Pbm	12·2 min	
^{200}Pb	21·5 h	EC, γ
^{201}Pb	9·4 h	EC, β, γ
^{202}Pb	3 × 10^5 years	EC
^{203}Pb	52·1 h	EC, γ
^{203}Pbm	6·1 s	
^{204}Pb	stable	
^{204}Pbm	66·9 min	
^{205}Pb	3 × 10^7 years	EC
^{206}Pb (Ra G)	stable	
^{207}Pb (Ac D)	stable	
^{208}Pb (Th D)	stable	
^{209}Pb	3·3 h	β
^{210}Pb (Ra D)	20·4 years	β γ
^{211}Pb (Ac B)	36·1min	β γ
^{212}Pb (Rh B)	10·6 h	β γ
^{213}Pb	10·2 min	γ B
^{214}Pb (Ra B)	26·8 min	β γ

m = metastable.   EC = electron capture

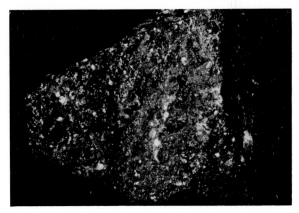

Figure 3.1 Galena, the lead sulphide ore.

Table 3.3 Lead–containing minerals

Galena	PbS	Anglesite	Pb SO_4
Bournonite	Cu Pb Sb S_3	Plumbojarosite	Pb $Fe_6(SO_4)_4$ $(OH)_{12}$
Boulangerite	Pb_5 Sb_4 S_{11}	Beudantite	[Pb, Fe, S (AsO_3)]
Jamesonite	Pb_4 Fe Sb_6 S_{14}	Linarite	(Pb Cu)$_2$ SO_4 $(OH)_2$
Meneghinite	Pb_{13} Sb_7 S_{23}	Leadhillite	Pb_4 SO_4 $(CO_3)_2$ $(OH)_2$
Cannizzarite	Pb_3 Bi_5 S_{11}	Lanarkite	Pb_2 SO_5
Sartorite	Pb As_2 S_4		
Jordanite	Pb_{14} As_7 S_{24}	Crocoite	Pb Cr O_4
		Vauquelinite	[Pb, Cu, CrO_4 PO_4]
Cotunnite	Pb Cl_2	Fornacite	[Pb, Cu, CrO_4, AsO_4]
Nadorite	Pb Sb O_2 Cl	Stolzite	Pb WO_4
		Wulfenite	Pb Mo O_4
Curite	Pb_2 U_5 O_{17} $4H_2O$		
Gummite	[Pb, U, Th, OH]	Descloizite	Pb (Zn, Cu) VO_4OH
		Pyromorphite	Pb_5 Cl $(PO_4)_3$
Cerussite	Pb CO_3	Mimetite	Pb_5 Cl$(AsO_4)_3$
		(Campylite)	
Phosgenite	Pb_2 Cl_2 CO_3	Vanadinite	Pb_5 Cl $(VO_4)_3$

Subsequently this lead ore is smelted, first by sintering to remove sulphu and finally in a furnace to produce lead metal. At this stage the lead contains many impurities, including antimony, arsenic, bismuth, copper gold, silver, tin, zinc and traces of other metals. Generally these metals are removed by electrolytic processes. Silver and gold are removed by the **Parkes process**, in which molten zinc is used to dissolve these metals.

History

Lead is one of the seven metals which have been known to man from earliest times, the others being gold, silver, mercury, zinc, copper and tin. One of the earliest references to it is found in Egyptian hieroglyphics of about 1500 BC, and it was found both as lead plate and statues in the tomb of Rameses III. In the Old Testament, lead was amongst the spoil taken by the Israelites from the Midianites (Numbers, ch. 31, v. 22), and Ezekiel (ch. 7, v. 12) described lead as an item of trade with the Phoenicians. It was mined extensively both by Greeks and Romans; indeed, Hadrian's wall is thought to have been built partly to protect the mines in Northumberland and Cumberland used by the Romans at that time.

In Roman and pre-Roman times, some of the importance of lead was in its association with silver; when silver-bearing ore was obtained, lead was removed as litharge by absorption into bone ash to leave the silver behind. Indeed, recovery of silver was so important that lead mines were often called silver mines. As much as 17 kg/tonne (600 oz/ton) of silver-bearing ore were obtained in mines in Asia Minor, although little silver was found in British mines. Roman lead technology was impressive, and the metal was used largely to make aqueducts and water mains. These aqueducts were lined with lead, and such lining has been preserved in many places in Roman Europe. Lead pipes were made in those days by bending lead sheet round a wooden former, and large water reservoirs were often constructed by making lead-lined tanks— a practice which was followed by the Greeks, who also collected rainwater from lead-lined roofs and stored olive oil in lead-lined vessels. In all Mediterranean cultures lead was frequently used in building work. In stone buildings the lead was poured between joints and, where stones were cramped together with iron bars, the iron bars were held in the stone by lead.

Recognition of the toxic effects of lead post-dates practical applications. The first fully documented reference to lead colic was by Hippocrates (370 BC) but a very much fuller, and still factually accurate, description of the symptoms of lead poisoning was given by Nicander in the second century BC, and was a well-recognized syndrome when described by Dioscorides in BC:

The drinking of lead causes oppression to the stomach, belly and intestines with wringing pains; it suppresses the urine, while the body swells and acquires an unsightly leaden hue.

Dioscorides: *De Materia Medica*

The possible hazard associated with the use of lead piping in water systems was recognized as long ago as the first century BC by Vitruvius, and it has even been suggested that the fall of the Roman Empire might be ascribed to the use of lead acetate to sweeten wine. Citois and De la Mark in the seventeenth century recognized this practice as dangerous. Sir George

Baker in 1767 traced a lead colic called the "Devonshire colic" to cider which had been contaminated by lead, and he also gave in his treatise a vivid account of earlier Continental epidemics of lead poisoning. In 1814 Orfila published a treatise on toxicology in which lead was mentioned. The signs and symptoms of lead poisoning were therefore well recognized by the eighteenth century. However, legislation was not introduced to control these industrial hazards until the Factory Act of 1864, further Acts being introduced in 1878 and 1883; lead poisoning became a notifiable disease in 1899.

Lead has a considerable number of uses in the present technological society. It is used as a metal in many functions, ranging from sheets for roofing, to pipes and blocks for screening from radioactive emissions. It is used as an alloy in electric battery manufacture, in solders and bearings for modern internal-combustion engines; and in its compounds, such as alkyl leads for anti-knock properties in motor fuels, lead oxides and lead naphthenates in paints, and lead arsenates in insecticides. The versatility of lead is not due to a single property, but to a series of characteristics where its softness, malleability, density, corrosion resistance, and chemical and electrical properties all play some part. Table 3.4 lists some of these uses.

Table 3.4 Principal uses of lead

Grid structure and compounds in batteries.
Coverings for power and telephone cables.
Sheet and pipe in plumbing.
Sheet and castings for shielding against gamma-rays and X-rays.
Foil for packaging and damp protection.
Caulking.
Solders.
Shot and bullets.
A principal or secondary ingredient in bearing metals.
Type metals.
For plating, dipping or spraying on steel and other metals.
For improving machinability of steel.
Lead tetra alkyls in petrol.
Litharge and red lead in glass and glazes.
Pigments for paints and inks.
Frits for enamelling aluminium.
Sodium plumbite in oil refining.
In accelerators in the rubber, artificial leather and linoleum industries.
Lead arsenate in insecticides.
Compounds in plastics, matches, explosives.
Sound and vibration attenuation.
Lead naphthenate as a dryer in paints and linseed-oil products.

Human exposure

Exposure of man to lead and the pathways that this might take are shown in figure 3.2. Interest has grown in recent years in the problem of health

ffects in man of environmental exposure to lead. Since 1900 there has been
a dramatic drop in the number of deaths from lead poisoning, and indeed
this has been paralleled by the drop in the number of notifications of
ndustrial lead poisoning to the Factory Inspectorate (figure 3.3). For these
reasons, industrial lead poisoning and therefore industrial lead exposure
has become of less importance. In adults, lead poisoning (plumbism or
saturnism) is almost exclusively a function of such industrial exposure,
although it may also occur where illicitly-distilled whisky ("moonshine") is
condensed in lead pipes or soldered radiators, or stored in improperly
glazed earthenware; where old battery casings are used as fuel; and (more
unusually) where lead and opium pills are used improperly or not all pellets
are removed from shotgun wounds.

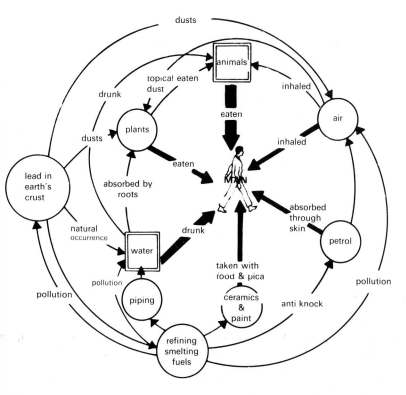

Figure 3.2 Routes of exposure of man to lead in the biosphere. The principal exposure is from
food, although secondary absorption may occur through other sources such as the air.

The possible health effects of smaller continual exposures to lead from the
environment are now recognized as being more important. In man, the

G

Figure 3.3 Notifications of industrial lead poisoning 1900—1970.

principal source of exposure is gastrointestinal absorption both from food eaten and fluid drunk. There are, in addition, absorption processes taking place through the lungs from lead in the air, and in very much smaller quantities through the skin. In the adult, about 10 per cent of ingested lead is absorbed from the gut, whilst it has been estimated that about 40 per cent of pulmonary lead (dependent on particle size and other factors) is normally absorbed. After absorption, lead is transported by the blood to the soft tissues, the highest concentrations being found in the liver and kidneys. Some, however, goes to other organs such as the brain, lung, spleen and heart. Deposition of lead takes place in bone (where more than 90 per cent of the total body lead is found) as relatively insoluble lead phosphates. Bone biopsies are difficult to obtain, but recent studies have shown that exfoliated teeth give a good measure of total lead exposure. All calcified tissues give a measure of cumulative lead exposure, but tooth lead concentrations are particularly useful, since it has been shown that lead may not be removed from teeth by chelation. Zonal analysis of teeth demonstrates that lead is concentrated in a circumpulpal secondary dentine zone, which is formed continually during the functional life of the tooth (figure 3.4). Measurement of tooth lead has therefore been used to monitor exposure in urban and suburban populations, and even to measure the lead burden in Egyptian mummies of the first and second millenia, and twelfth-century Peruvian Indian skeletons.

Most ingested lead is not absorbed, but is excreted in the faeces. The absorbed portion is excreted mainly in urine; small amounts are found in sweat and some is excreted back into the bowel (figure 3.5). In

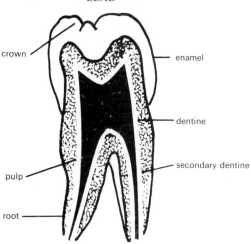

Figure 3.4 The distribution of lead in the tooth is such that the greatest concentrations are found in the circumpulpal dentine.

circumstances where there is alteration of the acid base metabolism of the body, lead may be mobilized from the bone and pass into the soft tissues. Thus, tissue exposure to lead may be two-fold: firstly in the acute phase of lead absorption, and secondly after lead mobilization from bone.

Recent work has indicated that lead absorption from the intestine of newborn animals and humans is very much greater than that in adults, and for this reason young animals and children are thought to be at particular risk when lead exposure is high. A question posed, and still remaining incompletely answered, is whether the evidence of increasing lead accumulation, especially in congested urban areas, is leading to an uptake of lead by humans sufficient to give rise to slowly-evolving and long-lasting adverse effects. The increase in lead in the biosphere has in part been confirmed by work on glacial concentrations of lead taken both in Greenland and Poland. The results indicated that a rapid rise in lead in the glacial ice commenced at about the start of the Industrial Revolution (1750), and thereafter rose until about 1940, when the rise became very rapid. The recent rise may be due to the increase in the use of lead alkyls as anti-knock agents in petrol. However, other studies have shown that the brain lead concentrations in Nubia about 1500 BC are not significantly different from those found in modern Danes.

In humans, many factors can alter the rate of absorption of lead from the gut; these are of considerable importance, since it would appear fairly certain that little damage can take place to any organism until such time as lead has been absorbed. Many studies have shown that changes in the

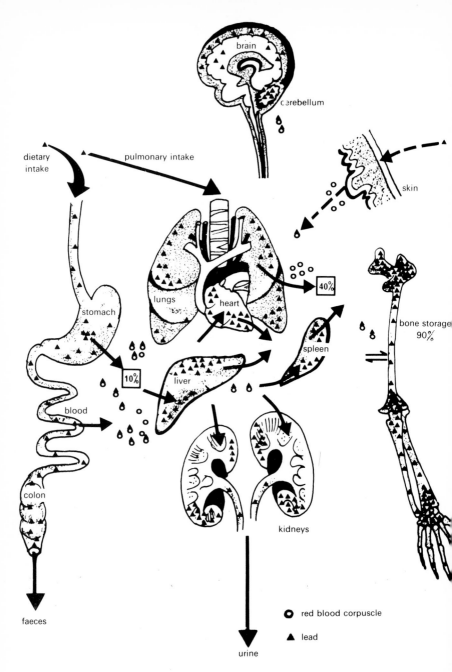

Figure 3.5 Uptake and distribution of lead in the body. Principal sites of intake are the gastrointestinal tract and the lungs. After absorption, lead is transported by the blood to soft-tissue sites and to storage sites in the bones. Lead is excreted in urine and faeces.

oncentration of calcium in the diet can alter the rate of lead absorption in in inverse manner. Similarly, alterations in protein and fat concentrations n the diet have been shown to alter the rate of uptake of lead.

Sources of lead exposure

1) *Food.* The uptake of lead from food would appear on average to be well within the upper recommended limit of 3 mg weekly proposed by the FAO/WHO Joint Committee on Food Additives. The average weekly intake of lead in food in Great Britain has been estimated at 1.2 mg. The intake of lead in children is about half that figure, although it would vary with age. Bryce-Smith and Waldron recommended the introduction of low-level lead foods for children and pregnant women, who are at greater risk. There is also some danger to health from contaminated seafoods caught in coastal waters, because of discharged sewage and industrial waste.

2) *Water.* It has now been appreciated that contamination of domestic water supplies can be a significant source of lead exposure, particularly in soft-water areas where plumbing systems are made of lead. In these areas it has been shown that there is a direct correlation between the level of lead in water and blood lead concentrations and, as will be explained later, these

section of lead waterpipe.

x 6

Figure 3.6 Section of lead pipe in use for about 100 years. This shows the oxidized and fissured interior surface.

findings have been associated with changes in the cardiovascular and nervous systems. These facts are especially true where water is stored in lead-lined tanks, although not exclusively so, since lead uptake into water has also been demonstrated from lead-soldered joints in copper piping where excessive quantities of flux have been used, It has been estimated from studies on old tenement buildings in Glasgow that no less than 6 mg of lead are ingested weekly in such a house, a figure greatly in excess of the upper recommended limit of 3 mg. The extent of the hazard in soft-water areas is clearly seen not only in the quantities of lead dissolved in the water, but also from the appearance of the interior of the lead pipes in such a situation (figure 3.6).

(3) *Air*. Intake from the air would appear to be only a minor component of lead exposure in Great Britain, but may be of greater importance in other countries. Studies have shown that a rise of 1 $\mu g/m^3$ of lead in the air can cause a rise of 1 $\mu g/100$ ml (0·05 $\mu mol/litre$) in the blood. Since the lead concentrations in air in congested streets are no greater than about 6 $\mu g/m^3$, this source of hazard would appear to be low, although it can become of great importance in industrial circumstances, such as smelting factories or where workers are using oxy-acetylene torches to cut up lead-paint-covered iron beams. Lead in the air can also be an important exposure factor where an organic compound such as tetra-ethyl lead is involved. This form can also be absorbed percutaneously, which may rapidly couse acute poisoning.

Recommended limits of lead intake from food, water and air are shown in Table 3.5.

Table 3.5 Recommended limits of lead intake

	Food (mg/week)	Drinking water lead concentration ($\mu g/litre$)	Air lead concentration ($\mu g/m^3$)	
World Health Organisation	3	100	—	
			Residential	High traffic density
European Economic Community (Proposed Directive)	—	50	2	8

Lead in soil and plants
The concentration of lead in soil can vary; the mean concentration is of the order of 10 mg/kg of soil. The figure is normally highest in the upper layers

of the soil, but is raised near lead smelters. Plants grown in such soils show increased concentrations of lead in their roots, although there is little translocation of lead within the plant. The principal exposure of lead in plants is that found topically upon the leaves next to busy main roads. Only a proportion of the total lead content of a soil is available to the plant. Soils with low organic content tend to promote uptake, whilst soils with high organic content bind lead within the organic component of the soil, making it unavailable to the plant. Only a small proportion of the lead absorbed by the roots is transported to the shoots to become available to grazing animals, although atmospheric lead may settle on the aerial parts of the plant and in this manner be eaten. When lead does enter the body of these animals, most of it will pass into the bone, and will therefore not constitute an exposure hazard to humans.

Lead poisoning in animals

Lead poisoning is probably the most common cause of accidental poisoning in wild and domestic animals because of their habit of chewing or licking objects indiscriminately. Thus, dogs and cattle can often develop plumbism whilst chewing posts or other wooden items which have been painted with lead-based paints. Wildfowl are of particular interest because they are hunted with shotguns which use lead pellets; when not killed, they can subsequently develop poisoning because of the pellets inside them or secondarily develop poisoning through ingestion of pellets during their search for food. Fish are not exempt from these effects of lead, and it has been shown that fish can be killed by water which has passed through old lead-mine areas.

Lead poisoning in man

Although lead has been used for 6000 years, there is still speculation and controversy about its possible harmful effects. There is no uncertainty about the illness produced by exposure to large amounts, but there is doubt about the possible harmful effects it may have upon the health of communities exposed to small amounts over a long time. It is most easily considered if first divided into different types as shown.

Clinical lead poisoning — acute
                       — chronic

Subclinical or biochemical lead poisoning.

Clinical lead poisoning. By this we mean that collection of symptoms experienced by the patient, and signs seen by his doctor, whose development is associated with exposure. When applied to disease, the terms *acute* and *chronic* refer solely to the time-course of the illness. In lead poisoning, the application of these terms has become rather imprecise in recent years, and it is certainly true that there is no clear division between the two, but rather a gradual spectrum. At the extremes, however, the forms of the illness are quite distinct.

Acute lead poisoning. The ingestion, in fairly large quantities, of lead salts such as lead acetate, produces immediate severe illness with burning pain in the mouth, throat and stomach. This is followed by severe colicky abdominal pain and constipation or diarrhoea, often with the passage of blood, because of severe irritation of the gut lining and, presumably, spasm of its muscle wall. In severe poisoning there is then failure of the heart and circulation, the kidneys and the liver, the patient finally lapsing into coma and death.

A contribution to the fall in incidence of this type of poisoning has been made by changing medical practice. Although heavy metals have long been recognized as potentially harmful to health, they were until recent years dispensed widely as medicaments. For example, until the introduction of antibiotics, arsenic was used in the treatment of syphilis, and is still used in treating some tropical diseases; mercury-containing diuretics were the mainstay in treating fluid over-load in conditions such as heart failure, and one of the common anti-diarrhoeal agents was lead-and-opium. Lead-and-opium tablets are no longer available, but shortly before they were withdrawn they were the cause of several unusual cases of poisoning in Glasgow. A small group of drug addicts stole some tablets from a chemist's shop and, having dissolved them, injected the material intravenously. There can never have been cases of lead poisoning better deserving the term *acute* than these. Their "fix" cost them dear, two members of the group dying soon after the injection.

This has not been the only occasion when toxic effects have resulted from intravenous lead, however, since the treatment of malignant disease by this method was in vogue for a time in the 1920s. The only remaining therapeutic use of lead at this time is in some dermatological preparations.

Chronic lead poisoning. This much more common type is a condition generally of less dramatic, more gradual onset produced by exposure to lead in the environment over a period of weeks, and more commonly months or years. The illness may come on gradually. The symptoms may be vague; perhaps a general feeling of unwellness with vague pains in the abdomen, trunk or limbs. If the patient is known to have contact with lead, perhaps in his work, the diagnosis may be simple, but sometimes the source of lead may be far from clear—a high lead content in the domestic water supply, for instance. In the absence of an apparent source of lead, other possibile diagnoses will be pursued, and the patient may be subjected to investigation over many months; even surgery has been undertaken for abdominal pain after a diagnosis of appendicitis had been made.

In some cases, although lead will have been accumulating over months, the onset of symptoms may be dramatically sudden. The diagnosis may be made immediately and treatment begun, with prompt relief; the condition then has all the characteristics of an acute illness.

Industrial lead exposure

They were the respectable ones. Well what did they get by their respectability? I'll tell you. One of them worked in a white lead factory twelve hours a day for nine shillings a week until she died of lead poisoning. She only expected to get her hands a little paralysed; but she died . . . that was worth being respectable for, wasn't it?

(Mrs. Warren's Profession, G. B. Shaw, 1902)

This quotation from George Bernard Shaw's play poignantly exemplified just what industrial lead poisoning meant at the turn of the century. In 1900, 1058 cases of lead poisoning were notified, of which 58 were fatal.

Industrial exposure is still the commonest cause of lead poisoning in adults. This danger has long been recognized in certain industries, such as lead smelting, paint manufacture and printing. In the United Kingdom, statutory health checks on workers in these industries have acted as a good early-warning system, and have been an influence in promoting the marked reduction in the incidence of poisoning from these sources. Death from lead poisoning in industry has now become so rare that it ceased to be a notifiable cause of death in 1966 (figure 3.7). Workers in industries such as

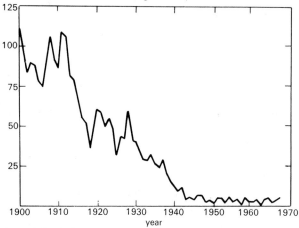

Figure 3.7 Annual deaths from lead poisoning in Great Britain, 1900–1966. After 1966, the disease is no longer classified separately as a cause of death.

scrap metal or demolition have contact with lead which is generally regarded as too slight or sporadic to warrant official monitoring. Several cases each year arise from such sources, however, and occurrence of just one case of lead poisoning draws attention to them.

"Moonshine" whisky. Many cases of lead poisoning arise, particularly in Kentucky and Tennessee, from drinking whisky made illegally, using old car radiators in distillation apparatus. Lead comes from the solder used as pipe connections. The nervous system seems to be particularly affected.

Domestic water. Poisoning has arisen from this source and, although seldom severe, one fatality was claimed in 1941, where the lead content of the domestic water supply was eighty times the present WHO limit. A high lead content of tap water was the only source of lead in several recent cases of lead poisoning in rural districts of Scotland.

Children, especially those living in a poor socio-economic environment, are at risk. This results from a combination of their displaying pica— that is, a habit of eating materials which should not be eaten—and the presence of old flaking lead-based paint in dilapidated properties. Such paint appears to have a rather pleasant sweet taste.

A common picture of lead poisoning in the adult
This is a man in his mid-thirties working in jobs involving oxy-acetylene metal cutting, often on demolishing buildings or bridges, scrap-metal work or shipbreaking; never long with one employer, he moves from job to job "chasing the money". He may be ignorant of the risk of lead exposure or, more commonly, take a buccaneering attitude towards it. He may elude the safety net of the Employment Medical Advisory Service and go to his own doctor complaining of stomach cramps and tiredness.

   A Victorian railway station in Glasgow (St. Enoch) was recently demolished. This involved cutting the steel girders coated with the paint of many decades, with oxy-acetylene burners. The paint was later found to have a lead content of 20% dry weight. Many of the men involved on the job had elevated blood lead levels, and ten of them had features considered severe enough to require their admission to hospital. One man was admitted to hospital as an emergency because of abdominal pain, and might have had an operation but for the knowledge that he was due to be admitted the following day for treatment of lead poisoning.

   The occurrence of the typical features of inorganic lead poisoning in this group is shown on figure 3.8. It will be seen that while anaemia and dysfunction of the gastrointestinal tract are common, serious effects on the

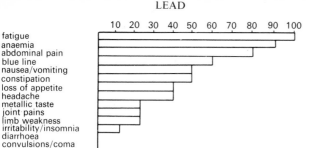

Figure 3.8 Incidence of the typical features of lead poisoning in ten oxy-acetylene "burners".

brain giving convulsions and coma are rare. The so-called "Burtonian" gingival blue line is found only when there is poor dental hygiene, when sulphur bacteria produce H_2S which combines with lead to form lead sulphide at the gingival margin.

Organic lead poisoning. Serious effects on the nervous system are more common in poisoning with organic compounds of lead. The best known of these are tetra ethyl (TEL) and, more recently, tetra methyl (TML) lead, the antiknock additives to petrol. Attention was first focused on the toxicity of TEL when, following its introduction in 1923, appreciable illness and even death was found among those involved in its manufacture. The nervous system was predominantly affected and, in the fatal cases, large concentrations of lead were found in the brain.

Neurological symptoms are still found to be prominent in this condition, even when the blood lead level is not greatly elevated; unfortunately, permanent intellectual damage may result. The affinity of organic lead compounds for lipid, a major constituent of the brain and central nervous system, may explain the difference from inorganic lead poisoning.

Cases of this condition are found where leaded petrol has been used other than as a fuel, for example as a solvent, or where men have been cleaning tanks which have contained petrol.

Presentation in children. Here too, the main concern is with the effect of lead on the brain, perhaps because the developing brain is more subject to insult by toxic agents. Abdominal pain is uncommon, but drowsiness and balancing difficulties may herald convulsions and coma with raised intra-cranial pressure. Even when death does not supervene, there is an alarmingly high incidence of mental retardation (up to 40%), with paralysis and visual disturbances thereafter.

Pathological effects of lead

It can be seen that lead poisoning gives rise to a wide range of symptoms. Some of these are quite readily explained by the known biochemical and patho-physiological effects of lead, but the basis of others is less clear. Lead is a general poison and indiscriminately affects all body tissues. Nevertheless, some tissues and organs are particularly affected.

The blood. Lead poisoning causes anaemia, that is, a reduction in the concentration in the blood of the oxygen carrier, haemoglobin. The way in which lead produces anaemia is probably complex. Three mechanisms seem to be involved:

> Suppression of haem synthesis
> Disturbance of globin synthesis
> Haemolysis, i.e. the premature destruction of the red cells in the blood which contain haemoglobin.

Almost certainly, the first is the most important. The interference by lead with the enzymes involved in the haem biosynthetic pathway will be discussed later. However, not only does lead reduce the incorporation of iron into protoporphyrin by suppressing the enzyme ferrochelatase, but it also renders iron less available.

Iron is able to enter the developing red cells in the bone marrow, indeed it is taken up more avidly than normal, but is unable to reach its site of utilization. It is trapped in inactive forms in the badly disrupted cytoplasmic organelles. Haem synthesis is therefore hindered, as it were, both by suppression of the means of production and by shortage of raw materials. Finally, the mere presence of so much unuseable iron in the cell seems itself to reduce haem synthesis.

Globin, the protein moiety of haemoglobin, is made up of α-and β-chains. *In vitro* experiments have shown that cells exposed to lead make less globin *in toto,* and also that the α- and β-chain production become out of phase. The suppression of α-chain production is more marked than that of β-chain. *In vivo* there appears to be a compensating mechanism which restores the balance, and so the significance of this finding is in some doubt.

The normal life span of a red cell is about 120 days. In haemolytic anaemia, the mean life span of red cells, as measured by radio-isotope techniques, is reduced. The fault may lie in the cells, making them age prematurely. These effects of lead produce changes in the microscopic appearances of the blood, which are characteristic of, although not unique to, lead poisoning. Basophilic stippling is the description given to the appearance of dark staining particles within the red cells. These are nucleoprotein remnants from the degenerating nucleus of the immature cell and indicate that, because of abnormal haem synthesis and haemolysis,

mmature red cells are being released from the bone marrow into the
peripheral blood. The white cells also show dark staining granules, called

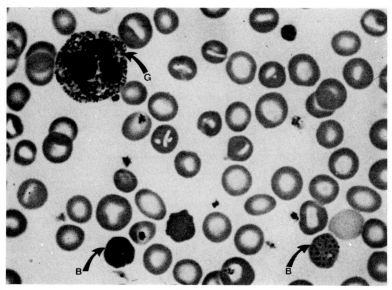

Figure 3.9 Blood film from a lead-poisoned subject showing stippled basophils (B) and toxic granulation in a leucocyte (G).

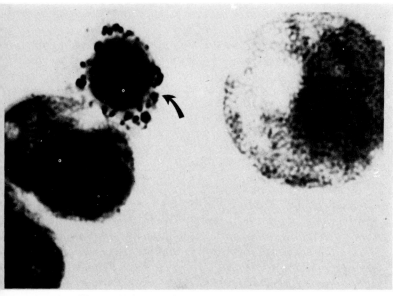

Figure 3.10 Marrow film showing a ring sideroblast.

toxic granulation (figure 3.9). Iron-packed cells called *sideroblasts* can be found in the marrow (figure 3.10).

The nervous system. This is made up of the central nervous system (the brain and spinal cord) and the peripheral nervous system. Lead poisoning seldom affects the brain very seriously in adults except, as has been said, in the case of organic lead poisoning. Lead-poisoned children are, however frequently affected; for example, by convulsions, paralysis and coma at the time of the attack, and intellectual impairment later.

The way in which this acute brain damage, known as *encephalopathy,* is produced is not clear. Fatal lead poisoning is so uncommon now compared with the start of the century, that opportunities for pathological investigation are few. The nerve cells of the brain are certainly damaged but this may not be the primary lesion. During life there is evidence of raised intracranial pressure. Examination of the brain in fatal cases shows it to be swollen, with the convolutions on its surface flattened by pressure against the inside of the skull. When sections are examined microscopically, it is found that there is an inflammatory reaction around the small blood vessels supplying the brain. This makes them more permeable to fluid, and it is this fluid leak which produces the swelling; this is possibly responsible for the convulsions and coma characteristic of severe poisoning.

Damage to the nerve cells themselves is probably not directly related to this increased pressure, since similar damage is found in cells of the peripheral nervous system which are not subject to increased pressure.

Peripheral nerves convey messages from the brain to the periphery especially muscles (motor neurones) and supply information to the brain (sensory neurones). In lead poisoning there is a reduction in the speed of conduction through some nerve fibres. For some reason motor neurones are particularly affected by lead, giving severe muscle weakness. Tracings made of the electrical activity in affected muscles (electromyograms) show a pattern characteristic of denervation.

The kidney. A description of the ultra-structural changes in the kidney in lead poisoning, which are best appreciated on electron microscopy, lie outside the scope of this discussion, but some of the effects will be considered. The kidneys are made up of units called *nephrons,* which consist of a filter system, the *glomerulus,* and the *tubules* through which the filtrate passes, both to have further material added by active excretion and to have filtered substances reabsorbed. The action of the glomerulus is passive, but excretion and reabsorption in the tubule allows delicate control of the body's biochemical balance.

It is this excretory mechanism which is mainly affected. Substances appear in the urine which would not normally be present, e.g. sugars and amino acids. On the other hand the tubular excretion of uric acid is markedly depressed. Uric acid is the substance produced by the breakdown of protein, and its accumulation in the blood causes its deposition in crystalline form in the joints, leading to the painful condition known as gout.

The glomeruli are subject to only minor acute changes. However, their number is greatly reduced in the chronic nephropathy which can result from prolonged exposure to high levels of lead, for example in moonshine whisky drinkers. Lead nephropathy was first documented in Australia between 1890 and 1930, when lead-based paints were extensively used on the exterior of houses. Autopsy studies showed that patients with chronic nephropathy of unknown cause had a higher content of lead in their bone than did patients with other forms of renal disease.

The blood, nervous system and kidneys are the parts of the body most affected by lead poisoning, but other organs can also be affected. Liver failure is said to accompany severe acute poisoning, but the nature of the damage has not been closely studied. The mitochondria, which may be regarded as the workhorses of the cell, have been shown to be damaged here as in other tissues.

Some abnormalities have been found in the electrocardiogram, but the heart's function is not generally impaired. The gastrointestinal tract is the seat of most of the common symptoms in lead poisoning— nausea, vomiting, abdominal pain, constipation and so on. We can only speculate, however, on the means by which these symptoms are produced. It may be an effect upon the autonomic nerves which supply the gut muscle, or it may be a direct effect on the muscle itself. Since bone is the seat of lead deposition, there is generally a gradual increase in the bone lead concentration with age. It is particularly found in the growing ends of long bones in children who are poisoned. This shows up as a line of radio-density on X-ray (figure 3.11). This is a fairly late finding. Lead does not seem to interfere with bone growth or integrity.

Many of the effects which have been discussed either partly or wholly recede following treatment. Tragically, the more long-term effects can be very serious. These include damage to the central nervous system, especially in children, resulting in mental retardation and occasionally paralysis or blindness, and lead nephropathy which may itself prove fatal in time, or may give high blood pressure leading to further kidney damage, or coronary artery disease or strokes.

Biochemical effects of lead

Lead has many diverse biochemical effects which are all of a deleterious

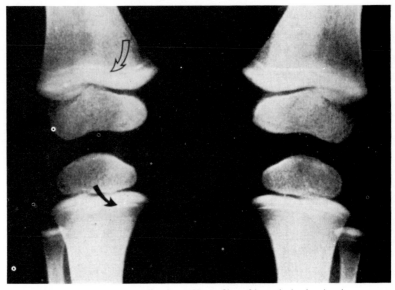

Figure 3.11 Radio-dense line in an X-ray film of bone in lead poisoning.

nature. No evidence has been presented of a possible essential function for
lead in metabolism. It has a valence shell electronic structure of $6s^2 6p^2$ and
normally exhibits a valency of two. The vacant 6p orbitals may be filled by
the formation of covalent bonds; this occurs most readily in biological
systems with sulphur atoms in proteins. These biochemical effects,
therefore all probably relate ultimately to the capacity of lead to combine
with specific biochemical ligands. These include sulphydryl groups, the
amino groups, carboxyl groups, phenoxy groups and imidazole residues. In
such circumstances lead might be expected to alter tertiary structures of
biochemical molecules, and therefore alter or destroy their biochemical
role. It may, in addition, substitute for essential metals in molecules, and in
such a position yet once again alter its tertiary structure. Consequently, lead
can change enzyme activity and destroy structure-function relationships of
nucleic acids.

Lead can, in addition, produce tissue damage. Since at the cellular level
the first structures to be exposed to the toxic effects of lead are cell
membranes, considerable change may be associated with their structure,
leading to release of tissue protein and increased extra-cellular enzyme
activity. For example, there is considerable morphological change in the
structure of the intestinal villi subsequent to exposure to only moderate
concentrations of lead acetate *in vivo* (figure 3.12). In circumstances where
lead exposure may have damaged cell membranes, increased plasma

(a) normal animal

(b) lead-exposed animal

Figure 3.12 Scanning electron-micrographs of rat small intestinal (jejunum) villi.

activities of the transaminases have been reported, and in addition lead inhibits $Na^+ K^+$ dependent ATPase which in erythrocytes results in a K^+ efflux.

H

It is clear from many studies that lead can markedly inhibit mitochondrial respiration. In addition to defective structure of their membranes, isolated mitochondria show impaired oxidative phosphoryl-ation, and ultimately mitochondrial respiration. Such inhibition of terminal oxidative pathways is not confined to animals, but has also been demonstrated in plants.

Of the complex of enzymes that synthesize succinyl Co-A from α-keto glutarate, the dithiol enzyme lipoamide dehydrogenase is markedly inhibited by lead. This inhibition is probably representative of the effects of lead on many enzymes either containing sulphydryl groups or being activated by sulphydryl compounds. Of particular interest in this sphere are the effects on haem biosynthesis (figure 3.13). Innumerable studies have shown that lead markedly inhibits both erythrocyte and hepatic ALA dehydratase activity in various species, including man (figure 3.14). Another enzyme within this same biosynthetic pathway which has also been shown to be inhibited by lead is ferrochelatase, the final enzyme of the pathway. Lead is bound preferentially to cellular organelles; thus, not only

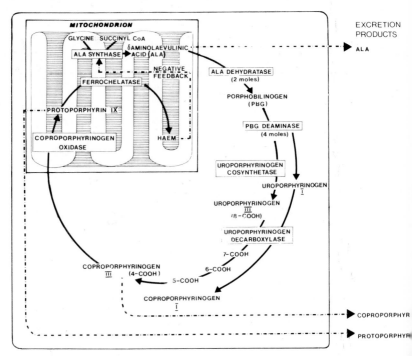

Figure 3.13 Haem biosynthesis in the cell; excretion products found to be elevated during lead poisoning are shown on the right. Typically, urinary and blood δ-aminolaevulinic acid, urinary coproporphyrin and blood protoporphyrin levels are found to be elevated.

are the functions of enzymes within the mitochondrion altered, but in addition there is clear evidence that fairly low levels of exposure can bring about marked changes in the morphological structure of the mitochondrion. The net result of these effects of lead upon haem biosynthesis is that large quantities of various precursors of the haem biosynthetic pathway are excreted in excess. Characteristically, excess quantities of delta-aminolaevulinic acid and coproporphyrin are found in the urine and, in addition, there are marked increases in the concentrations of zinc protoporphyrin in the erythrocyte. These rises in excretion levels of coproporphyrin are mediated through depressions of the mitochondrial enzyme coproporphyrinogen oxidase.

The elevation of delta-aminolaevulinic acid is mediated through two factors; first, through the depression of ALA dehydratase (figure 3.14) and secondly, through elevations of the initial and rate-limiting enzyme of haem biosynthesis, ALA synthase, by depression of free haem feedback inhibition. Studies have established that ALA can pass through the blood brain barrier of rats and rabbits at these concentrations and, upon injection, ALA can cause marked changes in the spontaneous activity of mice and rats. Indeed, in its own right ALA has various pharmacological activities. The possibility, therefore, exists that ALA can act directly or indirectly on the neuromuscular junction and cause some of the neurological effects of lead toxicity, although lead may well cause these

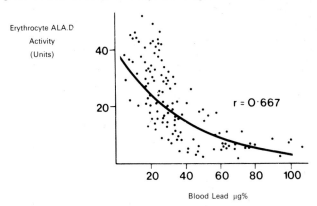

Figure 3.14 The depression of blood ALA dehydratase activity by lead in man.

effects through inhibition of important enzymes in this site, such as by inhibiting acetylcholinesterases.

Protein synthesis is markedly depressed by lead. Whether this effect occurs through alteration of the tertiary structure of RNA or inhibition of ability of RNA to bind amino acids is not known. It has, however, been shown that RNA from lead-intoxicated animals contains considerable quantities of lead, and that it causes disaggregation of polyribosomes in rabbit reticulocytes. Some alterations of protein synthesis may be due to alterations in the plasma concentrations of individual amino acids, thus inhibiting the rate at which protein synthesis may take place.

Although lead is concentrated by the nucleus, and can cause chromosomal abnormalities, there is no evidence to show that it can cause foetal abnormalities or cancer in man, although renal cancer has been found in rats intoxicated with lead. The chromosome damage by lead may be a function of release of DNase from liposomes which can damage chromosomes; DNase is markedly increased in urine in lead-poisoned animals.

The biochemical effects of lead, therefore, are probably almost exclusively a function of three different factors: first, the binding of lead to both functional and structural proteins, thus altering their functional or structural integrity; second, the alteration of the tertiary structure of nucleic acids, inhibiting the rate of protein synthesis; and, last, the displacement of essential metals from various biological molecules which therefore, by altering tertiary structure, again inhibit functional or structural activity.

Sub-clinical or biochemical lead poisoning

This is the name given to the concept that exposure to relatively small amounts of lead in the environment over a long time may have a detrimental effect on health in the absence of symptoms characteristic of lead poisoning, and in some circumstances without any perceptible change in the usual biochemical parameters. This is one aspect of the whole controversial field of study of pollutants in the general environment and their possible effects on health.

On general grounds, it is a fair hypothesis that a substance which is harmful when present in large amounts over a short period may also be harmful in smaller amounts over a longer period. That the effects produced in the first circumstance may not arise in the second, does not negate the argument. After all, one boxer may be knocked unconscious by a single punch with complete recovery. Another may be able to boast that in a long career he has never been knocked out, yet may retire with chronic brain damage; the first of these effects must have been common knowledge throughout human history; the second has come to be recognized only in very recent years.

It is not unusual for quite marked biochemical disturbances to be found in the absence of any symptoms of lead poisoning, as is often the case in industrial exposure. The measurement of blood lead is the most obvious test but has been shown to be somewhat inaccurate. Furthermore, there is

ome evidence that it is more closely related to contemporary exposure, i.e.
concentration in the environment, than to the lead load in the body. For
outine screening purposes, therefore, the use of various biochemical
ndices has been advocated. In general, these depend upon disturbances in
he haem biosynthetic pathway. The suppression of several enzymes in this
pathway as has been shown (page 86) causes the build-up of intermediate
products which either accumulate in the cells or spill over in excess in the
urine. Of these, red cell protoporphyrin and urinary δ-aminolaevulinic acid
and coproporphyrin are the parameters best related to lead poisoning.
Enzyme activity itself is altogether too sensitive to slight elevations in lead
level to be of monitoring value in industry. The haemoglobin level,
provided other causes of anaemia are excluded, is of value. A common
practice is to provide an initial assessment of an exposed subject by testing
blood lead concentrations and the haemoglobin.

Unlike those substances essential for life and health, there can be no
normal blood lead level. Any range quoted must be arbitrary; the generally
accepted figures are shown in Table 3.6. It may be asked why a higher level

Table 3.6 Normal range of lead concentrations

Blood or industrially	0–40 μg/100ml or 0–1·9 μmol/litre up to 80 μg/100ml or 3·9 μmol/litre
Urine	0–90 μg/24 h or 0–435 nmol/24h 0–65 μg/litre or 0–340 nmol/litre.

s acceptable in industrial workers. This is to some degree a counsel of
expediency, a recognition of the realities of the industrial situation. On the
other hand, there is variation in the reaction to lead among individuals, and
long-term workers in lead industries tend to be a self-selected group who
experience symptoms only at higher levels. There is understandable
pressure for these limits to be lowered, and certainly in the general
population any blood lead level over 2·0 μmol/litre (40 μg/100ml) is
unacceptable and should be investigated.

In the United States, it has been shown that people living in an urban
environment have higher blood lead levels than those in a rural
environment. A study of female black subjects showed that there was a
gradient of blood lead concentration away from a main road. On the other
hand, a fairly recent study of London taxi drivers showed no difference
between day-time and night-time drivers, despite significant differences in
exposure to exhaust fumes.

It is in children that the most damaging long-term effects of frank lead
poisoning occur; we have already referred to the chronic neurological
sequelae of frank lead poisoning—mental retardation, spastic paralysis
and so on. Thirty years ago a follow-up study of 20 children with "mild lead

poisoning" showed that only one subsequently progressed satisfactorily at school. This and similar studies relate, of course, to symptomatic lead poisoning, but they do stimulate interest in possible effects on intelligence and behaviour of levels of lead unassociated with symptoms and not generally regarded as toxic.

It has been shown that mentally retarded children have a higher blood lead concentration than normal children. This does not prove that it is the cause of the mental retardation, however, since there is a higher incidence of pica in retarded children. Nevertheless, several reports in the last two years have indicated neurological changes, behavioural abnormalities and evidence of intellectual impairment impairment in children whose blood lead was 2·0–3·5 μmol/litre (40–70 μg/100ml) compared with control groups.

Lead in the blood of pregnant mothers passes to the foetus. At high levels it may produce abortion or stillbirth, and this was known when women were subject to high industrial exposure last century. High levels have been found in the brains of stillborn infants. In soft-water areas a main source of lead is from the domestic water supply, especially where this is stored in lead tanks or conveyed in lead pipes. The foetus may be exposed through its mother's blood stream, and subsequently the young infant may be exposed to lead in bottled milk made up with water from this source. It has recently been shown that there is an increased risk of developing mental retardation in a child whose mother has been exposed to very high levels of lead in the water supply during pregnancy and in the first year of life. Although this study awaits confirmation, it is one of the first clear pieces of evidence of damage to health in the absence of a "toxic" blood lead level or classical symptoms of lead poisoning.

Although harmful effects on the developing brain of the young child are of principal concern, the possible place of lead in the aetiology of other diseases in the adult is also under consideration. This arises from the increased incidence of certain diseases, notably those of the heart and circulation, in areas where there is increased lead in the environment. In rats fed on a diet to which lead was added, ultrastructural changes developed in the heart muscle cells; this was accompanied by evidence of enzyme suppression of the cardiac muscle. It would be unwise to extrapolate the relevance of this experimental animal finding to the incidence of sudden death in human coronary artery disease. There may well be other factors present or absent in soft water which may explain these findings.

High blood pressure (hypertension) is known to be a factor in the production of coronary artery disease. One of the findings of a health survey programme in the West of Scotland is that subjects with hypertension have significantly higher blood lead than normal controls. Here again lead, at a level not usually regarded as toxic, has been shown to be associated with a disease process. In addition, renal insufficiency has also been associated with excessive lead exposure.

Prevention and treatment—prophylaxis

It is important to preserve a sense of proportion in considering environmental pollution. Certain facets of the problem require immediate attention. Outflow from industrial sources such as smelters should be carefully monitored, especially in areas of dense population, and the future siting of such industrial complexes should be carefully considered. Lead in the domestic water supply (mainly in soft-water areas, on present evidence, but perhaps also in hard-water areas) should be monitored to conform to acceptable levels. This will require attention to lead plumbing systems, especially where drinking water is stored in lead-lined tanks and, indeed, in the future, to the use of lead-soldered joints in copper systems.

Lead in petrol, and consequently in vehicle exhaust emissions, has probably received more public attention than any other source. Clear evidence that lead additives to petrol are detrimental to the health of the community is, however, lacking; but lead in exhaust emission must be contributing, along with industrial emissions, to the overall lead loading in the biosphere, and therefore should be limited.

Food is the principal source of lead intake. The present intake in foodstuffs in the United Kingdom does not exceed the levels suggested by WHO, but continued monitoring is essential, especially of tinned foodstuffs for infants.

Overall, wherever lead is presently used, a search should be maintained for effective, economically viable alternatives. Its removal from modern paints, already well advanced, and its removal or diminution from automobile fuels are examples of this. For workers in industry, especially women, there must be continued surveillance, and it is important that reasonable pressure for the "acceptable" degree of exposure for industrial workers should approximate more with that of the general population.

Treatment of lead poisoning

When frank lead poisoning develops, the first step should be the removal of the patient from the source. In mild cases, this may be all the action required. The need for further treatment is a matter for clinical judgment. In general, in the presence of a high level of lead, together with symptoms and biochemical evidence of an appreciable degree of poisoning (especially a depressed haemoglobin concentration in the blood), it is considered best to hasten the elimination of lead from the body by the use of chelating agents. These substances are so named because of their strong affinity for metal ions. Their molecular structures are shown in figure 3.15.

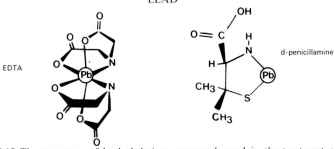

Figure 3.15 The structures of lead chelating compounds used in the treatment of lead poisoning.

The first chelate to be introduced (no longer used in the treatment of lead poisoning) was Dimercaprol. More effective, less toxic, substances were subsequently used; first the calcium disodium salt of ethylene diamino tetracetic acid and, more recently, penicillamine. This last has the advantage that it can be given orally with good effect.

Other measures include calcium salts for abdominal pain, very occasionally blood transfusion in severe anaemia, and physiotherapy for limb weakness resulting from neuropathy. Lead encephalopathy, with its potentially disastrous consequences, is a medical emergency, since everything possible must be done to relieve raised intracranial pressure.

There has been more published work on lead poisoning and lead exposure than on almost any other subject in toxicology. Any substance like lead which, because of its usefulness to man, has been used since antiquity and will continue to be used by modern and future societies, but which has deleterious effects on the animal organism, will always attract a considerable degree of interest. Recognition of the earliest toxic effects is a considerable step forward in the prophylaxis. Paracelsus said:

All substances are poisons: there is none which is not a poison. The right dose differentiates a poison and a remedy.

The problem of lead in the environment is to determine the level that is both acceptable to human economy and compatible with the health of the community.

FURTHER READING

Barltrop, D. and White, J. ed. (1975) "Lead Symposium," *Postgraduate Medical Journal*, **51**, 744.

Chisolm, J. J., Jr. (1971), "Lead Poisoning," *Scientific American* **224**, 15.

Goodman, L. S. and Gilman, A. (1970) *Heavy Metals: in Pharmacological Basis of Therapeutics*, 4th edition, Macmillan, London & Toronto.

Lead—Airborne Lead in Perspective (1972) National Academy of Sciences, Washington, DC.

Proceedings of International Symposium, Amsterdam 1972. "Environmental Health Aspects of Lead." Commission of the European Communities, Luxembourg, May 1973.

Survey of Lead in Food (1972), Ministry of Agriculture, Fisheries and Food, London HMSO.

Survey of Lead in Food—First Supplementary Report (1975), Ministry of Agriculture, Fisheries and Food, London. HMSO.

Waldron, H. A. and Stofen, D. (1974). *Subclinical Lead Poisoning*, Academic Press, London and New York.

CHAPTER FOUR

ARSENIC

M. D. KIPLING

Arsenic and its compounds

Arsenic (As), an element of mass number 75 and atomic number 33, is a member of the same group in the periodic table as nitrogen, phosphorus, antimony and bismuth. It is widely distributed and forms 5×10^{-4} per cent of the earth's crust.

Table 4.1 Some naturally-occurring compounds of arsenic

Mineral name	Formula
Arsenolite	As_2O_3
Arsenopyrite	FeAsS
Cobaltite	CoAsS
Cu_3AsS_4 Enargite	
Kupfernickel	NiAsS
Lollingite	$FeAs_2$
Mimetite	$Pb_5Cl(AsO_4)_3$
Mispickel	FeSAs
Niccolite	NiAs
Olivenite	Cu_2OHAsO_4
Orpiment	As_2S_3
Proustite	Ag_3AsS_2
Realgar	AsS
Scorodite	$FeAsO_42H_2O$
Smaltite	$CoAs_2$
Tennantite	$Cu_8As_2S_7$

The commonest naturally-occurring arsenical compound is mispickel, an arsenical pyrites, which occurs together with other metallic ores. Less

93

widely distributed are the two brightly coloured sulphide pigments, realgar which is orange red and orpiment which is of a yellow lemon colour. Arsenical sulphides are also combined with copper (enargite) and silver (proustite), and arsenic also occurs in some nickel, cobalt, silver and lead ores.

Arsenic is produced as a by-product from metal mines in Australia, Canada and Brazil, and has been extracted as a contaminant of lead and copper from smelting works in the United States. The largest known deposit is a sulphide deposit containing 10% arsenic at Boliden in Sweden, where the excess of arsenic produced was sufficient to require its disposal by sinking in concrete containers in deep water. In Britain, a compound of arsenic and tin is the most common ore. Almost all arsenic is now produced as a by-product of roasting other ores. The volatile oxides of the metal are deposited in the flues; on reheating, the material collected contains up to 90% arsenic. Further refining produces a product containing 99% As_2O_3. The small world requirement for metal arsenic is met by reducing arsenic trioxide (white arsenic) with charcoal in a retort, or heating arsenical compounds in the absence of air. In 1975 the leading producer was Sweden with about 15,400 tons. There were only four other major producers: Mexico at 9400 tons, France 9100, the Soviet Union 6600, and South West Africa (Namibia) about 6350.

Of the inorganic arsenic compounds, arsenious oxide (As_2O_3), known as white arsenic or simply as arsenic, is the most important. It is sparingly soluble in water to produce arsenious acid $HAsO_2$. It reacts with alkalis to form arsenites. Arseniuretted hydrogen (arsine gas, AsH_3) may be prepared in a pure condition by the action of dilute sulphuric acid upon sodium or zinc arsenide. A series of irritant liquids which are derivatives of arsine have been manufactured for warfare. These include methyldichlorarsine ($CH_3As\ Cl_2$) and its ethyl and phenyl homologues, as well as the toxic smokes, such as diphenylchlorarsine (DA), diphenylcyanarsine (DC), and the well-known war gas lewisite (chlorovinyldichlorarsine, $Cl\cdot CH = CHAsCl_2$). A number of important pharmacological substances have followed the manufacture of sodium p-amino phenylarsinate ($NH_2C_6H_4As\ O\ (OH)(ONa)$ or atoxyl). The discovery of its spirochaetal and trypanocidal properties led to the development of a number of less toxic and more effective arsenicals, such as arsphenamine and neoarsphenamine.

Tests for arsenic

There are three chemical methods used for the detection of arsenic. The first was described by Reinsch, who made use of the observation that metallic arsenic is deposited on copper plate when a solution of arsenious acid is in the presence of pure hydrochloric acid. The test involves the boiling of a sample of the suspected material in a container with arsenic-free copper foil. If arsenic is present, a grey deposit occurs on the copper; if dried and heated, the arsenic condenses as characteristic octahedral crystals.

The second test, Marsh's, utilizes the capacity of nascent hydrogen to reduce to arsine any arsenic or arsenic compounds present in the material. Hydrogen is prepared by the action of dilute sulphuric acid on arsenic-free granulated zinc. The suspect material is then added to the reaction vessel and the escaping gas ignited; any arsenic present will be deposited on a cool surface applied to the end of the flame. The arsenic deposited from the arsine on the tube is sublimed and the crystals identified.

Gutzeit's test is more sensitive and, like Marsh's, depends on the formation of arsine. In this test the arsenic reacts with silver nitrate or mercuric chloride or bromide on filter paper, any contaminating sulphuretted hydrogen having been removed with lead acetate. A positive test is shown by the development of a yellow-brown stain; the depth of the colour may be used to give a quantitative estimation.

A more recent and sensitive test for the detection of arsine uses silver and diethyldithiocarbonate in the presence of excess silver nitrate. In the presence of arsine, a red colour is produced and is compared with standard tubes to estimate the concentration of arsenic. A level of 0.025 parts per million is detectable by this method.

A most important advance in the identification and estimation of arsenic, particularly in biological specimens, was made by the development of activation analysis. The principle is that the arsenic to be measured is converted to its radioactive isotope[76]As, which emits beta and gamma radiations and has a half-life of 27 hours. In practice a small sample of the suspected material in a polythene bag is inserted in a nuclear reactor together with 1–2 mg of arsenious oxide for comparison. After irradiation the sample is digested in acid and 10 mg of arsenious acid added. This acid, together with the irradiated arsenic, is converted by the Gutzeit method to arsine and absorbed on mercuric chloride. A comparison can then be made with a Geiger or scintillation counter of the pulses received from this material, and the pulses received from the known amount of arsenious acid that has been similarly irradiated.

Arsenic in the environment

Arsenic is ubiquitous in the biosphere. It is present in concentrations from 1 ppm in limestone and siliceous deposits, 2 ppm in igneous rocks, and up to 20 ppm in volcanic rocks. The highest figures of up to 10,000 ppm have occurred in the Waitapu valley in New Zealand and Buns, Switzerland. It is oxidized by natural processes to the pentavalent condition, and it is in this state that it is found in soil. In virgin soil and forest humus, an arsenic content of 3–5 ppm has been found. In the sea the concentration is 3 ppm of the salt content, but marked variations are found in uncontaminated fresh water. Its presence may be detected in almost every sample, but significant amounts occur in freshwater springs with a high bicarbonate level, and in saline lakes. In the United States it has been estimated that there is an average daily intake of 10–12μg of arsenic from drinking water. High concentrations in well water have been found in Europe and the Argentine; these concentrations have been connected with disease both in animals and man.

Naturally-occurring arsenic in the environment is augmented by that extracted from the earth by man, and often converted from the pentavalent to the more active trivalent state. Arsenic has been found in the atmosphere from coal burning; one power plant which used a coal rich in arsenic emitted a ton daily. It is present in some degree in most coal in Britain, and elsewhere up to 8000 ppm has been found in coal ash. It is also present in the neighbourhood of smelters and industrial processes in which arsenic is used. It is added to soil by the use of phosphate fertilizers, arsenical pesticides and defoliants, though it is not necessarily absorbed by all plants when present in the soil in fertilizer. Sufficient arsenic was found to have accumulated in the soil of an orchard to render the soil subsequently infertile. The poor growth of barley has sometimes been connected with excess arsenic accumulated in the ground.

Arsenic in food

All foodstuffs contain small amounts of arsenic (Table 4.2). It occurs naturally in plants (and therefore in foodstuffs) almost always in the pentavalent form which has low toxicity. The arsenic of industry and commerce is almost always in the trivalent form which is much more toxic. The distinction is important, since many wholesome foodstuffs contain the pentavalent form at levels above the legal limit. This limit has, however, been prescribed to deal with contamination by trivalent arsenic compounds.

A very detailed study was made by a Royal Commission in 1903, which was appointed after a severe outbreak of arsenical poisoning in beer drinkers in the Midlands and North of England. The Commission reported that much of the arsenic found in foods could be attributed to the use of sulphuric acid made from arsenical pyrites in the manufacture of sugar, citric and other acids, vinegar, glycerine, yeast and food colourants. They found that another cause of contamination was the use of coke containing arsenic. When this coke was used for drying malt, the arsenic impurity affected the beer. Other foodstuffs such as grain and vegetables have also

Table 4.2 Arsenic in foodstuffs

	ppm
Haddock	2.2
Oysters	2.9
Clams	3.1
Beef	1.3
Pork Liver	1.1
Lamb Chop	0.4
Whole Wheat	0.2
Rice	0.1
Rhubarb	0.5
Mushrooms	2.9
Cocoa	0.6
Tea	0.9
Table Salt	2.7
Sugar	0.1
Butter	0.2

een found to be contaminated, and arsenic from pesticides has been found
many foods. Lead arsenate as a protection from caterpillars was used on
uit and was a particular hazard in apples from the United States. The
senic content was reduced by weathering, but more efficiently by the
troduction of washing with dilute hydrochloric acid and then water.
alcium, manganese and copper arsenates are other possible con-
minants.

The spraying of vines does not lead to contamination of all wine, because
e arsenic is precipitated in the lees by the yeast, but fruit juices which have
ot undergone fermentation retain a higher proportion. Green vegetables
ay contain excess arsenic but potatoes, although they may be sprayed do
ot absorb significant amounts from the leaves into the tubers. Cocoa
eans have contained arsenic from processing. Contamination of gelatin
as the cause of sufficient concern to give rise to the enactment in Great
ritain of a statutory limit under the Edible Gelatin (Control) Order 1947.
rsenic compounds were frequently found in some food preservatives and
olourants, including iron pigments made with contaminated sulphuric
id, in wrapping papers and in shellac, which was an early form of plastic
ed as a lining in vats and casks.

rsenic in man

nce arsenic is present in air, food and water, it is continually absorbed by
outh, the lungs, and possibly through the skin. The total daily arsenic
take has been estimated to be of the order of 400–1000 μg. The
stribution in the body (Table 4.3) depends on the form of arsenic
gested; arsenites have a particular affinity for the hair, nails and skin.
rsenic has been found to enter the hair roots within 30 minutes of
gestion.

Table 4.3   Arsenic in body tissues (dried)
(data from Liebscher, K. and Smith, H. (1968) *Arch. Environ. Health.* **17**, 882.

Whole blood	0.04
Bone	0.05
Brain	0.01
Hair	0.46
Kidney	0.03
Liver	0.03
Lung	0.08
Nail	0.28
Skin	0.08
Thyroid	0.04

Since arsenic is present in even the unexposed subject in amounts great
than that of some essential trace elements (such as iodine and cobalt) the
is a possibility that like these elements it is a necessary constituent of th
human body; this has not so far been proved. Arsenic in small doses is sai
to improve the appearance and increase the vitality of some mammals. A
arsenic material found on rocks in the Austrian Alps was traditional
eaten to increase the strength and endurance of mountaineers and t
improve the complexion of women. Sir Thomas Oliver, the famou
occupational physician, wrote that it made the men more lively, combativ
and salacious, and the women more comely, with the result that there we
an inordinate number of illegitimate children there.

The uses of arsenic

Historical

It is certain that arsenic compounds were observed by man from the earlie
times. The naturally-occurring pigments would draw attention by the
bright colours, and the melting of copper and tin ores in the Copper ai
Bronze Age cultures produced detectable deposits of arsenic oxid
Arsenical pigments were used for decoration in Egyptian tombs, ai
arsenic was mined and used as a dye by the Persians, from whom it deriv
its name *zarnic* or *arsenicon* in Greek. In classical times the authors of bo
medical and scientific works described various arsenical materials.

Pliny the Elder (who died when observing the eruption of Vesuvius
AD79) advised the local application of arsenic in the treatment of ulcer
and Celsus, a great writer on medicine in the reign of Tiberius, advised c
its prescription. His writings received little attention until, through tl
influence of Pope Nicholas V, they were published in 1478 in Florence

one of the first printed medical textbooks. Galen, at the end of the second century, advised the use of arsenic, as did Dioscorides Padanius, a surgeon in Nero's army, who wrote a popular book on medicine that was copied at regular intervals.

After the fall of the Roman Empire, medicine and science were kept alive in Europe by the Arabs in Spain. Maimonides, a Jewish physician, advised as a precaution against arsenic poisoning that unnaturally coloured food be given to others to taste before eating. Avicenna, the great physician of that world, also described the use of orpiment as a medicine. The Arab chemists, such as Rhazes and Geber, worked with arsenic and to the latter is attributed the discovery of arsenious acid which he obtained from the tri-sulphide. It has been called *white arsenic* from that day to this. It is also said he knew of a metallic arsenic and that it could be produced from white arsenic as a deposit on copper. The element is thought to have been first prepared in 1250 by Albertus Magnus by heating the oxide with soap, while Brandt, an eighteenth-century chemist, demonstrated that white arsenic was the calx or dephlogisticated form of the metal, as oxides were then described. Albertus Magus was later canonized, and is perhaps unique in being both a chemist and a saint.

The knowledge gained of the properties of arsenic in man led to its widespread use as a poison. There was a law in the Italian town of Siena, passed in 1365, which limited the sale of arsenic sulphide to the owner of a pharmacy; it forbade his dispensing it to any slave, servant, child or youth under the age of 20, and laid down that it should be given only to adults who were well known. A similar Act was passed in 1851 in Great Britain. Early evidence of the use of arsenic as a poison is recorded in a trial of a minstrel in Paris in 1384, who was alleged to have been asked by Charles the Bad, King of Navarre, to poison his brother, the King of France, two uncles, and other members of the nobility at court. The arsenic was collected from a number of apothecaries in the city to be sprinkled in the soup and other food at the court. Arsenic appeared in the English language in the fourteenth century as "Arsenicum Hyghte Auripigmentum for the colour of golde and is gadereyd in Pontus"; Chaucer quotes its use.

By the sixteenth century, arsenic was well known as *ratsbane,* and the poisoning of enemies by arsenic became common. The use of arsenic preparations as a medicine was widespread and *auripigmentum* (yellow arsenic and sandarax) appears in a Medical Dispensatory of 1608. In the seventeenth century arsenic was used in a cake applied to the breast for the treatment of plague, as an ointment to be injected into tuberculous ulcers (Kings Evil), as an external and internal treatment for cancer, and in the form of Belmonts Balsam was highly recommended for venereal ulcers. The Edinburgh New Dispensatory of 1788 devotes many pages to its properties, bringing the recently published work of Dr Fowler of Stafford to the notice of its readers, for the treatment of agues and headaches.

Many other preparations were fashionable, and the value given to arsenic as a medicine was such that the Admiralty required ships with crews of 40 and above to carry a supply of Liquor Arsenicalis (Fowler's solution, containing potassium arsenite). Arsenical preparations were considered to be of value in treating cancer, and in the nineteenth century the list of diseases for which arsenic was a remedy gives an indication of how widespread its use had become. The list included debility, anaemia, epilepsy, asthma and chronic skin diseases. Fowler's solution continued to be prescribed for leukaemia and disseminated sclerosis until the middle of the present century, and is still used to a very small extent for some intractable skin diseases. Orpiment is still used in some countries as a depilatory.

The organic arsenicals came into prominence when it was discovered at the Liverpool School of Tropical Medicine that atoxyl (p. 94) was an effective drug in animals experimentally injected with trypanosomes, the cause of the tropical disease of trypanosomiasis or sleepy sickness. This finding was the greatest benefit that arsenic has brought to mankind, because the German scientist Paul Ehrlich, who had experimented on the use of dyes as bactericides, showed that atoxyl would kill the spirochaete of syphilis. Arsenicals had been used both by mouth and injection since the introduction of syphilis into Europe in the fifteenth century but, owing to the superiority of mercury by mouth and the local and general toxicity by injection, they had been little used. Arsenic compounds were known to kill spirochaetes *in vitro,* but it was Ehrlich who, by synthesizing hundreds of organic compounds, produced 606 arsphenamine and the improved 914 neo-arsphenamine, which were toxic in the body to spirochaetes but relatively non-toxic to man.

The discovery of the nature of atoxyl in 1905 and Ehrlich's researches ensured that the very large number of cases of syphilis after the 1914-1918 war could be given effective treatment; in 1920 there were over 40,000 new cases in England and Wales. A single course at that time would entail the injection of 20–30 g of arsphenamine and generally had to be repeated. The introduction of penicillin as the treatment of syphilis means that Tryparsamide is the only organic arsenical at present used for injection. It is of particular value because it can penetrate the blood-brain barrier. It is also used in advanced sleepy sickness. Acetarsol, a derivative of arsanilic acid (Ehrlich 594) was until recently available as a pessary for protozoal infections of the vagina.

Contacts with arsenic

The most important sources of contact with arsenic in the nineteenth century were arsenical pigments. The coloured substances in everyday use included all types of paper, food wrappings, toys, fabrics and cloths, rubber articles, paints, soap and sweets; the manufacture of these pigments employed a substantial number of people. Workers were also exposed in the manufacture of sheepdip and other pesticides, glass, anti-fouling paints, and a variety of chemicals. Arsenic was used in the manufacture of lead shot, Britannia metal, copper and brass bearings, the copper of locomotive boiler tubes, and some platinum alloys. Arsine gas, which until recently has only existed through accidental generation or in laboratory testing, is now used in the manufacture of semi-conductors.

In agriculture, the arsenic compounds French Green and Paris Green were used as insecticides in France at the beginning of the nineteenth century, but because of their toxicity their use was made illegal in 1846. However, their value was such that they were quickly re-introduced; but at the end of the century they were replaced by lead arsenate and calcium arsenate.

Arsenic compounds, especially sodium arsenite, have been used to kill vegetation, for removing bark and killing tree stumps, and for suppressing weed growth in reservoirs. Arsenic compounds were of great economic value in cotton and tobacco growing, and lead arsenate is still used in vineyards and apple orchards. Sodium arsenite was found to be the most effective agent against chewing insects such as the cotton boll-weevil and the apple codling moth; in Britain it is the only arsenical on the list of Approved Products for Farmers and Growers, and is recommended for the control of winter and small ermine moths (apples and pears), pear slug sawfly (cherries), earthworms and leatherjackets (turf). In animal husbandry arsenic preparations are used as larvicides in poultry droppings, as a treatment for enteritis in turkeys, and for chest diseases, anaemia and skin diseases in horses and cattle. In show dogs arsenic has used as a tonic to improve the general health and skin condition. Arsenious oxide is still in use as a rat poison. Lead arsenate has also been employed for regulating the growth of grapefruit and derivatives of phenylarsonic acid, such as arsanilic acid and sodium aminarsonate, for growth stimulation in poultry and pigs.

Arsenic usage in industry and agriculture has recently declined very sharply; it has been replaced by the modern range of insecticides and herbicides and by changes in industrial technology. Arsenic trioxide imports to the United States fell from approximately 22,000 to 13,000 tons between 1968 and 1971.

I

Table 4.4 Arsenic compounds used in industry, agriculture and medicine

Arsenic compound	Formula	Known as	Uses
Arsenic	As		Alloying additive
			Electronic devices, i.e. transistor, etc.
			Veterinary medicines
Arsenic pentoxide	As_2O_5	Arsenic oxide	Chemical intermediate
		Boliden salts	Defoliant
			Wood preservative
Arsenic trioxide	As_2O_3	Arsenic	Insecticides and fungicides
		Arsenolite	Glass
		White arsenic	Chemicals
		Arsenious oxide	Anti-fouling paints
			Taxidermy
			Timber preservation
Arsenic trichloride	$AsCl_3$	Butter of arsenic	Pharmaceuticals and chemicals
Arsine	AsH_3		Stabilizing selenium in transistors
Calcium arsenate	$Ca_3(AsO_4)_2$		Insecticide, herbicide and larvicide
Copper arsenite	$CuHAsO_3$	Scheele's green	
Copper aceto-arsenite	$3CuOAs_2O_3Cu$ $(OOCCH_3)$	Paris Green Emerald Green	Larvicide
Orpiment	As_2S_3		Depilatory— fireworks—pigment
Potassium arsenate	KH_2AsO_4	Macquer's salt	Preservation of hides
			Textile printing
			Fly papers
Potassium arsenite	$KH(AsO_2)_2$	Fowler's solution	Veterinary medicine
Realgar	As_2S_2		Pigment
			Depilatory
Lead arsenate	$PbHAsO_4$		Insecticide, herbicide, and growth regulator
Sodium arsenate	Na_2HAsO_4 Na_3AsO_4	Wolman salts	Wood preservative
			Calico printing
			Insecticide
			Weedkiller
Sodium arsenite	$NaAsO_2$		Herbicides
			Pesticides
			Corrosive inhibitor
			Chemical intermediate
			Fluorescent lamps
Magnesium arsenate	$Mg_3(AsO_4)_2$	Atoxyl	Trypanicide
Sodium arsanilate	$NH_2C_6H_4AsO$ $(OH)(ONa)$		Pharmaceutical manu- facture

Arsenic poisoning

Effects of arsenic

Excessive absorption of arsenic compounds (particularly the trivalen
form) leads to poisoning, and in suffcient quantities to death. A large dos

is followed after a short period by a feeling of constriction of the throat, difficulty in swallowing, and stomach pains. The pains increase, and severe vomiting and diarrhoea ensue. Acute shock, with cold clammy skin, a weak pulse and breathing, may be the prelude to an immediate death or death from exhaustion in the following few days. The symptoms are attributed to a direct irritant effect on the gut, and a toxic action on the muscle of the heart. Chronic arsenic poisoning is ushered in by a general weakness, nausea and loss of appetite, with possibly some vomiting and diarrhoea.

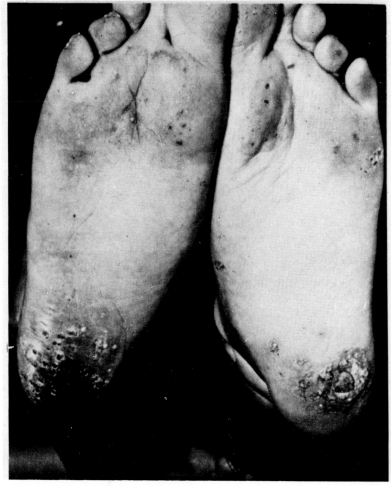

Dr. J. G. Holmes

Figure 4.1 An arsenic malignancy or pre-malignancy due to the administration of Fowler's solution for a case of disseminated sclerosis.

General catarrh follows with hoarseness and mild bronchitis. Later general lassitude and sometimes jaundice supervene.

A well-known symptom of chronic arsenic poisoning is the onset of peripheral neuritis. The nerves in the limbs are affected, with loss of sensation in the feet and hands, perhaps preceded by tingling. The sense of touch is diminished, and the cornea of the eye becomes anaesthetic and easily inflamed and ulcerated. The muscular power is also affected in severe cases, but the loss of sensation in the limbs may lead to difficulties in walking and controlled movements. The jaundice may progress or be followed by a fibrous degeneration of the liver and cirrhosis. The effect on the liver is typically the production of a cirrhotic or "hob-nail" liver, which can also be caused by alcohol.

In recent years another type of liver disease has been identified in which the pressure in the portal veins of the liver is increased, leading to back pressure, haemorrhages and enlargement of the spleen. In a recent case the patient, aged 35, had taken Fowler's solution daily for 22 years. He also developed skin changes, skin tumours (figure 4.1) and cancer of the larynx and probably the lung. Circulatory disorders in the limbs which may be followed by gangrene have also been reported.

Vineyard workers appear to have been especially prone to circulatory disorders; in one study of 180 German vineyard workers suffering from signs of arsenic poisoning, 6 developed gangrene, and 20 showed deficient circulation. There may be a local effect of arsenic on the skin, causing inflammatory and eczematous changes, but arsenic incorporated in the tissues of the hair, nails and skin produces more specific effects such as melanosis—a blackening and mottling of the skin, particularly the eyelids, neck, nipples and armpits, and in severe cases over the abdomen, back and chest. The hair may become brittle and fall out, and ridges occur on the nails.

The organic compounds of arsenic have been used on a vast scale for the treatment of syphilis, and to a lesser extent for tropical protozoal diseases such as trypanosomiasis. Intravenous injection is occasionally followed by immediate anaphylactic shock or, within 24 hours, by headache, fever, vomiting and diarrhoea. Later reactions include a wide range of disorders of the kidney, blood, skin and brain.

Poisoning by the inhalation of arsine is a dramatic event. The inhaled arsine liberates haemoglobin (from the red blood cells) which blocks the kidneys and the liver. The symptoms are pain in the loins and general collapse. The released blood pigments produce a bright red urine, a bronze colour of the skin and jaundice. In less severe cases, anaemia, and sometimes neuritis, prolongs the period before complete recovery.

Causes of poisoning

Poisoning has occurred in the past due to the handling of playing cards or bank notes, and the wearing of pigmented fabrics. The eating of arsenical pigments in bread contaminated on the shelves, in the coloured decoration of a blancmange and in coloured confectionery have also caused illness, as has the use of arsenical compounds for colouring sweet wrappers and children's toys. Further exposure would have occurred from similarly pigmented wallpapers, Venetian blinds, carpets, linoleum and book-bindings.

The medical profession drew attention to these hazards in official publications and the popular Victorian novelist Seton Merriman wrote a thriller *Roden's Corner* describing a secret arsenic pigment factory into which, in the guise of a charitable organization, workers were lured to ill-health and death. The first medical description in Great Britain was in 1831 by Charles Thackrah of Leeds and *Poisoning in the Manufacture of Arsenical Pigments* was published as an Official Report in 1863. As late as 1908 Dr E. L. Collis, a distinguished Medical Inspector of Factories reported:

> Hardly a single person presented for examination had a sound nasal septum. Ulceration and sloughing of the septum may occur as quickly as three weeks after commencing work, which may not be associated with any marked inconvenience to the patient, but is usually accompanied by a stuffiness of the head, resembling a severe head cold, and by headaches more or less severe; improving on separation of the slough. Earache, pointing to catarrhal otitis media may accompany the process of ulceration. In several cases this condition had progressed further, until deafness and ear discharge—suppurative otitis media—had been set up. Externally, there were signs of local irritation, sores on the skin, especially in the folds of the groins. There were also signs that general constitutional damage was caused, pigmentation of the skin was frequent, while several workers showed signs of peripheral neuritis.

An unusual pigment poisoning occurred more recently in the case of Mrs Claire Luce, United States Ambassador to Italy. The embassy had spacious seventeenth-century bedrooms, the beams of which were painted green; many coats of paint had been brushed into the rosettes to make them stand out against the background. Mrs Luce used her bedroom as a quiet retreat for dictation and writing, but after a year she found that her duties were more and more of an effort. She felt tired and ill, and suffered from nervousness and nausea. At an Art Festival in Venice, during a waltz she found that her right foot had lost its feeling. She was at first diagnosed as suffering from anaemia, and after treatment in hospital in the United States, she returned fit. Soon the symptoms reappeared and, in addition, her finger nails became brittle, her blonde hair was falling out and she became an irritable invalid. She was admitted to a Naval Hospital in Naples where, because of the inflammation of her mouth, tests for arsenic were carried out, which proved positive. After intense secret-service activity, political poisoning was ruled out and further investigation suggested the cause was the decoration of her bedroom where flakes c paint were liable to fall from the vibrations in the ceiling when anyon walked overhead. The fact that she had noticed her coffee, which she mad herself, tasted bitter and metallic was confirmatory evidence. It was finall considered to be the lead arsenate in the paint contaminating the room an her coffee, and perhaps the production of an arsine derivative by mould that was the cause of her ill-health.

The white powder of arsenious oxide has led to fatal errors. It has bee mistaken for flour (to be added to oyster soup), calcium sulphate (fc peppermints), and as a dust for sweets and cough drops. Not long ag arsenical contamination of rice caused over 100 acute cases of poisoning i orphanages in Sumatra. The consumption of a health food made from kel led to excessive arsenic absorption in two patients attending a neurolog clinic, though arsenic was not, after investigation, thought to t responsible for the symptoms. Poisoning has also occurred from ove eating of black molasses, and from contamination of sausage meat, but th most important and memorable outbreak was due to contamination c beer.

It was noted, at the beginning of this century, in the Manchester are and to a lesser degree around Liverpool and the north of Birmingham, tha the number of cases of alcoholic neuritis was rapidly increasin Alcoholism and neuritis were not uncommon conditions in the dispensarie and hospitals at that time, and the increase in this disease was at firs attributed to such an unworthy cause as spending the money on drink tha had been collected for soldiers in the Boer War. Dr E. S Reynolds of th Manchester Royal Infirmary recognized that the outbreak could be due t poisoning from arsenic. By detailed inquiries it was soon found that th arsenic was present in the beer from only a few brewers, and a commo factor in the brewing of these beers was the use of glucose from one firn This firm used an arsenic-contaminated pyrites in the production of th sulphuric acid in its manufacture. It is thought that about 6000 people wei affected with this arsenic poisoning, and at least 70 died. A Roya Commission was appointed and the minutes of this Commission publishe in 1903 demonstrated the extreme thoroughness of the investigation whic had been undertaken. Arsenic was shown to be present in the beer fro glucose, but smaller quantities were also found in the malt. The Roya Commission not only investigated, but recommended action which led t the laying down of definite limits of arsenic content in the products the had investigated and in foodstuffs generally.

The relevance of the discovery of excess arsenic in the body to the caus of death must often have appeared obscure in an age when many medicine contained arsenic. It is perhaps best illustrated in the case of Napoleon, the cause of whose death is not certain, and whose supporters consider that his British conquerors or French Royalists poisoned him.

Even within the last 20 years the widespread and unsuspected distribution of arsenic in materials has led to analytical errors. Following an outbreak of arsine poisoning, the slag and dross were examined for any arsenic contamination in order to find the source. Unexpectedly high results were obtained which were shown not to be due to contaminated reagents. Eventually it was found that the glass balls used to prevent bumping during the reactions were affected by the fluorides present in the slag. which released the arsenic they contained from the mix used in their manufacture.

Industrial poisoning

Industry has not been a major cause of arsenic poisoning in Great Britain, except in the manufacture of pigments. Glass making was noted to be prejudicial to health in the early nineteenth century; cases of poisoning occurred in 1902, and recently attention has been drawn to the risks in Siena. General ill-health with chronic chest and skin diseases was attributed to work in arsenic reduction works in Cornwall in 1903, and an official report was made to the Home Secretary on the excessive number of workers admitted to the neighbouring poor law hospitals.

The manufacture of lead shot was thought to be the cause of disease in two workers, but lead absorption may have played a part in the same way as occurs in lead arsenate exposure. One such worker aged 32 was thought to be sufferring from both lead and arsenic poisoning. He had been employed for one month in handling trays used for drying the precipitate from the action of arsenic acid on lead acetate. The grinding in a mill and filling of drums was a dusty occupation. The worker suffered from vomiting, constipation, giddiness and weakness of the limbs. There was some jaundice and anaemia, and both lead and arsenic were found in excess in the urine.

In sheepdip factories, sodium arsenite has caused both diseases and death. One worker aged 30, after nine days' exposure filling casks in very dusty conditions, was treated in hospital for sores on the skin and swollen eyes; within a few days of returning to work he developed intense vomiting and two days later he died.

Arsenic trichloride, which has caused death when absorbed through the skin, is a particular hazard and is a cause of acute poisoning. Some poisoning occurred in war-time conditions in the manufacture of organic arsenicals, but otherwise this has been carried out without incident. The notifications of arsenic poisoning to the Factory Inspectorate is an indication of the extent of industrial poisoning (Table 4.5).

Table 4.5 Notification of arsenic poisoning (excluding arsine): number of fatal cases in brackets

Process or occupation	1900–1918	1919–1939	1940–1959	1960–1974
Manufacture of arsenical colour	53(1)	4(1)	1	—
Manufacture of sheep dip	4	14(5)	8(2)	1
Refining arsenious oxide, etc.	10	4	4(1)	—
Manufacture of various arsenites	—	4(2)	—	—
Manufacture of various arsenates	—	4	2	
Arsenious chloride	18(1)	1(1)	—	—
Tannery depilatory	1(1)	2	—	—
Taxidermy	3	—	—	—
Manufacture of wallpaper	1	—	—	—
Manufacture of anti-fouling paint	2	—	—	—
Lead shot manufacture	1	1	—	—
Miscellaneous or information lacking	12	11	7	—
	105(2)	45(9)	22(3)	1

The British records, owing to the statutory requirement for notification are the most comprehensive that are available, but poisoning has been reported elsewhere. In Sweden, the smelting works at Rönnskär gave rise to *Rönnskär disease*. A survey in 1955 showed that among 1500 workers, infections of the throat and bronchitis were prevalent due to arsenic trioxide in the atmosphere.

The agricultural use of arsenic in Europe has been the most frequently described hazard, in particular in vineyard workers. Disease of the liver is a recognized complication of treatment with Fowler's solution and has occurred in industry and in the outbreaks of poisoning among beer drinkers. An article in 1774, "De hydrope ex Ingestio Arsenico Observatio", first drew attention to the problem, and it was further studied by Sir Jonathan Hutchinson, who in 1888 described general liver fibrosis in a young man. Since that date there have been regular reports, but the most striking examples are accounts of the disease among vineyard workers. In one report, enlargements of the liver, noted in 73 out of 113 workers, was attributed to arsenic, though consumption of alcohol may have had a part to play. In the Rhine and Moselle districts of Germany arsenical insecticides have been banned since 1942. Arsenic poisoning has occurred in fruit growers, gardeners and sprayers of arsenic on to potato crops. An estimate in the United States suggested that there were 50 deaths a year between 1946 and 1955 in agriculture from arsenic poisoning.

Water pollution

Contamination of water supplies by agricultural pesticides has been a further source of poisonings, though natural water was a cause of chronic arsenical poisoning in Taiwan and the Argentine. In the former outbreak, nearly 20% of 37,000 villagers showed evidence of arsenic posioning and in the latter the water contained up to 0.45% arsenic. Greater concentrations are found in some waters polluted by smelting works. At Reichenstein in Silesia, the works has given its name to "Reichenstein disease" in which ulcers in the mouth, digestive complaints, neuritis and skin changes are the common symptoms.

The latest accounts to be published describe the effects on a family of drinking from a well in Belgium polluted from a neighbouring factory. The father, aged 69, developed typical skin changes, and one son developed neuritis. Another son died of acute poisoning.

In Western Minnesota 11 out of 13 employees of a building contractor were poisoned from a well on the premises. Abdominal pain, diarrhoea and vomiting occurred, and in three cases polyneuritis developed. The well water contained up to 21 ppm of arsenic which was derived from an abandoned store of grasshopper bait.

Homicidal

Arsenic was recognized in history and in legend as a poison from the earliest times. Arsenic poisoning for homicidal purposes was particularly common in the eighteenth century. The accounts of many British trials, such as those of Madeleine Smith, Mrs Maybrick, Armstrong and Greenwood have become classics, but small numbers still occur. In 1949 white arsenic with sugar, mixed in pies, was used in a murder. The most recent case is the Stoneleigh murder, which illustrates the points of interest in deducing that death had occurred from arsenical poisoning. On 8th September 1969 Mrs Waite, wife of Lord Leigh's chauffeur, died of acute gastro-enteritis with allergic polyneuropathy. The family doctor who gave the death certificate later telephoned Dr Derek Barrowcliff, the Home Office pathologist, to ask his advice on why this disease was fatal in a previously healthy woman of 45 years. A search of the medical records showed that Mrs Waite had suffered from diarrhoea and vomiting, peripheral neuritis, scaly skin, swelling of the legs, loss of hair and appetite, abdominal pain, mental irritation and shingles—an occasional concomitant of arsenic poisoning. These are the complete list of symptoms of arsenic poisoning but, although a computer would have immediately made the diagnosis, it did not occur to the 20 doctors who examined Mrs Waite on teaching rounds in the hospital where she was treated for a month. At the post mortem, 40 ppm of arsenic was found in the liver, 150 ppm in the duodenum and 332 ppm in the small intestine. There could be no doubt that she died of arsenic poisoning. After excluding possibilities that arsenic in the hair was due to contamination by vomit or hair lacquer, Dr Barrowcliff was able to plot the concentrations of arsenic in the hair against the events of the illness (figure 4.2), showing clearly the low figures during

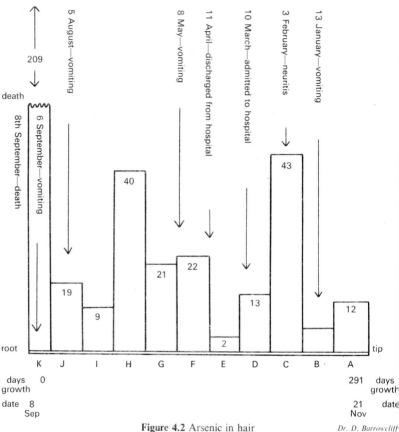

Figure 4.2 Arsenic in hair   *Dr. D. Barrowcliff*

her stay in hospital. The behaviour of the husband in attempting suicide when police inquiries were made , and requiring restoratives when a post mortem was requested, strongly supported the medical evidence.

Arsenic for homicidal purposes has the advantage of comparative tastelessness, but the disadvantage of comparative insolubility. It is difficult to administer in water, but milk, tea and cocoa are suitable vehicles. Large doses may be given in porridge without detection by the victim. Arsenic was readily available in the past from flypapers, which could be soaked in water to produce a lethal dose. In recent times arsenic was readily available as a weed killer, and this has been the source of the poison in several cases.

Besides administration for homicidal purposes, arsenic may be administered accidentally. The fatal dose is said to be in the region of 120–180 mg, although this dose may lead to immediate vomiting and therefore subsequent recovery. The actual effect of ingestion depends on the rate of absorption and method of administration. A husband was sentenced to death in 1816 for inserting arsenic as an abortifacient into the vagina of his wife, who subsequently died. A fatality also occured when a woman with acute infection of the vagina was treated by inserting 18 acetarsol pessaries.

Arsine poisoning in industry
Arsine poisoning appears in industry in the guise of an unexpected and severe gassing accident. The workers are unaware of any risk until they develop shivering, headache, vomiting and other symptoms of a general illness, followed by the passage of red urine. It would appear to be a rare condition, but it is likely that many cases, owing to an absence of the knowledge of the risk, are undiagnosed. In a recent case, a young man was employed for a short time on the traditional process of dipping brass articles into a liquid containing arsenic trichloride and hydrochloric acid to produce a bronzing effect. He was asked to try the effect of dipping a batch of zinc alloy door hinges; this resulted in the evolution of arsine. He was taken ill and admitted to hospital where, by elimination, a diagnosis of possible arsine poisoning was reached. Following inquiries from the hospital, investigation at the factory revealed that exposures had occurred, though neither the patient or the management had been aware of the possibility.

More frequently it is only the occurrence of simultaneous cases that points to a common industrial exposure. A typical example occurred in a British slag-washing plant. A 54-year-old worker was admitted to an infectious diseases unit with a diagnosis of infectious hepatitis. He had felt unwell when working the previous day and returned to bed in the afternoon; later he developed vomiting, abdominal pains and backache. Blood appeared in his urine and he became jaundiced. The possibility of an occupational cause was suggested when he said that his brother-in-law, who had worked with him, developed the same symptoms on the same day.

A visit to the factory showed that six other men had been employed for some years on a plant washing the fluorspar slag from an aluminium recovery furnace to extract a residue of aluminium. The slag was rotated in a drum, through which water from a local stream was directed. No arsenic was present in the process, but the provisional diagnosis of arsine poisoning was confirmed when it was found that three other workers had at some time in the past suffered from symptoms that could be attributed to arsine. In one case the blood in the urine had suggested a bladder tumour, in another the jaundice had been diagnosed as infective hepatitis, in another the anaemia had led to a diagnosis of a toxic blood disease, and in the fourth, since neuritis was also present, to a diagnosis of pernicious anaemia. It was only when two cases simultaneously attended the same hospital that the cause became apparent. In this particular plant, the source of the arsenic was obscure until it was discovered that at infrequent intervals a copper melt was treated which included scrap boiler tubes containing 0.36% arsenic.

Arsine was discovered by Scheele in 1775 and the first known fatality was a research chemist in Munich in 1815 who "inspired a small portion and at the termination of one hour was seized with continual vomiting, shivering and weakness which increased until the ninth day when he died".

The first reported industrial poisoning was in 1873 among workers desilvering lead and zinc ores; later there were three deaths among nine men treating silver ores with hydrochloric acid. In Great Britain the first identified cases were in 1894 in two workers recovering the ammonium chloride that is used in the surface of galvanizing baths, one fatal case treating spelter with impure hydrochloric acid, and five cases (of whom three died) in a process of dissolving zinc in hydrochloric acid.

The occurrence of arsine poisoning was fully described in 1908 by John Glaister, Professor of Forensic Medicine at Glasgow University, who collected 120 recorded cases, which he classified as follows:

I. Chemical operations in laboratories	
(a) Operations with known arseniferous materials	8
(b) Operations with unknown arseniferous materials	14
(c) Operations unknown	1
II. Trade processes	73
III.* Military ballooning	16
IV. Domestic environment (wall papers)	6
V. Causes unknown	2
TOTAL CASES	120

*The balloons were filled with hydrogen generated by the action of acid on zinc, in which arsenic is a common impurity.

He found that a number of occupations were involved, including chemists, physicists and physiologists

Our knowledge of the incidence of arsine poisoning in Great Britain is obtained from the statutory notifications since 1896 to HM Chief Inspector of Factories (Table 4.6). The information is not, however, comprehensive because many occupations were not covered by the provisions of the Factories Act; its requirements for notification were not always well known to all doctors and a significant proportion of cases would not have been diagnosed.

Table 4.6 Incidence of arsine poisoning by process or occupation (number of fatal cases in brackets)

Compound or process	1900–1918	1919–1939	1940–1959	1960–1974
Manufacture of various zinc salts	13(2)	15(3)	1	—
Cadmium recovery and preparation	—	4	3	2
Tin refining	—	13(2)	6(2)	—
Bismuth preparation	2(1)	—	—	—
Purification of zinc	2(1)	—	—	—
Electrolyte recovery of copper	3	—	—	—
Non-ferrous metals and drosses (miscellaneous)	—	4(1)	2	5
Manufacture of hydrochloric acid	3(2)	—	—	—
Paper industry	2	1	—	—
Manufacture of benzidine and other dyestuff intermediates	5(2)	1(1)	—	—
Chemical processes not otherwise classified	4(2)	2	—	—
Art bronzing	3(1)	1(1)	—	—
Pickling and galvanizing	8	—	—	—
Manufacture of silicon steel	—	—	2	—
Entry into vessels containing acid residues where not otherwise classified)	9(2)	—	4(2)	—
Miscellaneous	—	2(2)	5	10

The causes of accidental production of arsine may be divided into its evolution following the reduction of arsenical compounds and the hydrolysis of a metallic arsenide. The former includes such processing as galvanizing, cadmium recovery and the manufacture of zinc salts. Episodes investigated by Dr. A. T. Doig, HM Medical Inspector of Factories in Scotland, illustrated the often unrecognized danger of entry into confined spaces that have contained mineral acids. A steel scraper, an iron shovel, a galvanized bucket, and a copper shovel were used to remove deposits of sludge from containers that had contained sulphuric acid made from iron pyrites with a resultant evolution of arsine. The hydrolysis of metal arsenides has frequently occurred and been the cause of arsine poisoning, and the refining of tin has been a cause of serious outbreaks of poisoning. The arsenic impurity in tin is retained in the slag after smelting, and there is

sufficient tin in the slag to make further refining economic when the arsenic
is removed with the dross. To allay dust, water may be added and, to
improve the reaction of dross formation, aluminium is also added.

Aluminium has a special role in the production of arsine, either from
hydrolysis following the formation of an aluminium arsenide, or from the
evolution of nascent hydrogen from an electrolytic action. The presence of
water from a heavy thunderstorm acting on dross in a printer's foundry was
the cause of the first outbreak of arsine poisoning in this manner in Britain
but sufficient moisture may be present in the atmosphere or in sawdust used
on the dross to promote the evolution of arsine.

A serious incident in 1975 illustrates this hazard. In a small workplace a
new technique of crushing dross was in action, by which the dross was
treated in a ball mill, and air was blown to separate the constituents. The
separation of zinc from some dross was undertaken, and 10 tons were
satisfactorily processed. Two hours after further processing had started
the operators felt faint and weak, suffering from abdominal pains and
vomiting. The eight men concerned were immediately admitted to hospital
and were found to be suffering from arsine poisoning. They were treated by
exchange blood transfusion and renal dialysis. One of the cases was fatal.
The cause of the evolution of arsine was the wetting of the dross in
transport to the workplace, and it was found that there was aluminium
present in the dross.

Ships and poisoning

Arsenic and arsine poisoning have been connected particularly with ships.
The Royal Commission on Arsenic Poisoning in 1903 considered possible
cases studied by Major Ronald Ross of the Liverpool School of Tropical
Medicine (the discoverer of the cause of malaria). He described how in
ships' crews, who in those days were liable to beri-beri, there could well be
confusion in the symptoms of this vitamin deficiency and arsenic
poisoning. The sailors on long voyages in sailing ships, who were fed on
monotonous farinaceous diets, were deprived of vitamin B1 and developed
the pigmentation and weakness typical of the disease. This closely
resembled arsenical poisoning, which also occurred in ships through eating
rice treated with arsenical pesticides in the East.

The first suspicion of arsine poisoning at sea was in the transport of
ferro-silicon in the nineteenth century, but it is possible, as the symptoms
were not typical, that phosphine was mainly responsible for the casualties. A
dramatic example was found to have occurred on a Swedish boat when she
docked in 1907 in Antwerp. Of the six passengers, four were found to be
dead. It was at first assumed that since one was a fugitive from Russia

(whence he had sought refuge in Sweden) the Russian police had placed poison gas bombs in his cabin. It was found on investigation that arsine had been produced from a cargo of ferro-silicon that had been broached in a storm. Arsenic was found in excess in the deceased. Recently in a ship transporting cylinders of arsine, eight of the crew were exposed to arsine from a leaking cylinder, and four suffered severely. The transport of arsine is uncommon but extremely hazardous.

Arsenic and cancer
Skin cancer
Arsenic and its compounds in the environment do not present any great hazard of poisoning. The extent of exposures has been ascertained, and the risks removed by technological changes and supervision. Accidental production of arsine is the sole remaining matter for serious concern. Whether arsenic is a carcinogen in man still remains to be fully investigated; opinions have changed for over a century.

The first accounts of cancer related to exposure were given by Paracelsus in 1531 and by Agricola in 1536 who described *mala metallorum* in the metal mines of Saxony and Bohemia. They were not aware that the fatal disease was cancer of the lung, but they noted the high incidence and mortality of lung disease, and the fact that the miners' wives had a succession of husbands.

The first attribution of carcinogenic powers to arsenic was by Dr J. Ayrton Paris, a Penzance physician, who wrote in 1820:

> It may be of interest to record an account of the pernicious influence of arsenical fumes upon organized human beings, as I have been enabled to ascertain in the copper-smelting works of Cornwall and Wales; this influence is very apparent in the condition both of the animals and vegetables in the vicinity; horses and cows commonly lose their hoofs, and the latter are often seen in the neighbouring pastures crawling on their rumps whilst milch cows, in addition to those miseries, are soon deprived of their milk . . . It deserves notice that the smelters are occasionally affected with a cancerous disease of the scrotum, similar to that which affects chimney sweeps . . . and it is singular that Stahl in describing the putrescent tendency in the bodies of those who die from this poison, mentions in particular the gangrenous appearance of these parts. It is a very extraordinary fact that previous to the establishment of the copper works in Cornwall, the marshes in their vicinity were continually exciting intermittent fever, whereas since that period a case of ague has not occurred in the neighbourhood; I have heard it remarked by the men in the works that the smoke kills all fevers.

The account of his observations has in general been rejected, because no similar outbreak has occurred elsewhere in the world, and the cattle disease appears to be more related to exposure to selenium than to any known compound of arsenic. There are, however, some reasons to doubt the prevailing scepticism. Paris was not only a country doctor. At the age of 23 he was physician to the Westminster Hospital and later became a Fellow of the Royal Society and President of the Royal College of Physicians.

Cancer of the scrotum is a condition that cannot be mistaken or misdiagnosed, and is almost invariably due to occupational or external causes. It is now a recognized result of arsenic absorption. It would seem unlikely that a physician of such distinction would be in error.

Sir Jonathan Hutchinson was the first to provide (in 1887) definite proof that the long continual ingestion of Fowler's solution could lead to the development of skin cancer. By 1946 there were 143 cases in the world's literature which, considering the large numbers who were treated with arsenical preparations, would suggest a very small and insignificant incidence. The number reported, however, is certainly a very small proportion of the actual number of cases because, in the period when arsenical medicines were in vogue, medical publications were not as numerous as at present, and specialist clinics were few in number.

Recent studies may give a more realistic indication of the incidence. In Germany in 1966 a study of 180 patients with skin disease who had received arsenic showed that 21% had developed skin cancer, and there appeared to be a dose relationship. As recently as 1975 one consultant dermatologist had seen in his clinic 4 cases of epithelioma due to Fowler's solution; multiple skin cancers, including one on the scrotum, were described in two drinkers of polluted water.

Skin cancer occurs from industrial exposure to arsenic compounds and is well documented throughout the world. Solitary cases were reported from a smelting works in Mexico, from a Japanese copper smelter and from a copper mine. The disease was also said to have occurred in workmen handling arsenical ores.

The most comprehensive information available is from the investigations in Britain, where arsenic poisoning in industry has been a notifiable disease since 1896. Two cases were recognized by 1913, and by 1922 Sir Thomas Legge of the Factory Inspectorate wrote, "arsenic is recognized as associated with slowly developing epitheliomatous ulceration." Twenty years previously he had examined a group of sheep-dip workers who showed pigmentation of the skin and keratoses, and he found both had developed in three of the workers. By 1957 a total of 14 cases of skin cancer were known among arsenic workers. Skin cancer also occurred in the manufacture of arsenical insecticides. The use of insecticides has led to skin cancer in agriculture and in gardening. The numbers so far known are small, except in the case of vineyard workers in whom, however, there is the complicating factor that they absorbed arsenic from drinking their own wines. Skin cancer attributed to arsenic in drinking water was first described in 1809. It was particularly marked in Reichenstein in Silesia where there was contamination from arsenical ores. In other areas, such as Cordoba in the Argentine, drinking wells in which the water had a high arsenic content were a cause of cancer of the skin in the local population. In one report 65 cases were described, augmented by a further 26 in 1938. The cancers were often multiple, occurring particularly on the limbs.

Internal cancer

Internal cancer in exposed workers was first considered to have occurred in the Schneeberg and Joachimstal miners in Central Europe. The metallic ores of these mines included arsenical compounds such as cobalt arsenide, realgar and orpiment; dust in the mines contained about 0.1% of arsenic. The death rate was estimated to have been up to 70% of the miners, with a total death rate of 40% from cancer of the lung. In the Joachimsthal silver mines in Bohemia, nickel, cobalt, bismuth and arsenical ores were later mined, as was uranium towards the end of the nineteenth century. The miners suffered from a disease similar to that of the Schneeberg miners, and in 1926 this was shown to be a cancer of the lung. The dust from arsenical ores was first considered to be the cause of the cancer, but it has now been suggested that radio-active products of uranium are the causative agent.

In Britain there was a high incidence of cancer of the lung in persons employed refining nickel ore by the nickel carbonyl process. The technique was introduced in 1902 into South Wales, where concentrated ore was imported from Canada. By 1959, 62 cases of cancer of the nasal sinuses were attributed to dust, and 131 cases of cancer of the lung also occurred. Full investigations in 1948 showed that an excessive mortality from both cancer of the lung and cancer of the nose was almost entirely confined to workers in this refining process; following these reports many improvements were undertaken in the factory, including the use of arsenic-free sulphuric acid and reduction of exposure to fumes and dust. It was thought that the removal of arsenic impurities by the substitution of the sulphuric acid was the cause of the disappearance of the cancers, but the occurrence of similar cancers connected with nickel smelting by other processes suggests that a nickel salt may have been responsible. The role of arsenic is supported by the occurrences of nasal cancer in sheep grazing near a German smelter, but it is also argued that, as the flock was an inbred strain of Hampshire sheep and other animals nearer the factory suffered poisoning, but not cancer, the nasal growths were of genetic origin.

Medical Inspectors of Factories in Great Britain observed that lung cancer had occurred in the manufacture of sheep dips. In 1945 the Medical Research Council undertook a full investigation at a sheepdip factory. All the deaths in the small isolated town over the previous years were investigated, and the causes of death of the 73 deceased workers between 1910 and 1943 were identified. Those causes were compared with those of deceased farm workers, general labourers, and shopkeepers in the same town, and the cancer rate in the factory workers was found to be twice as high as in the control group. The excess of cancer which was of skin and lung occurred amongst those workers with a proved exposure to arsenic dust. Medical examinations of the workers at the time of the survey revealed the typical arsenical skin changes of pigmentation and thickened patches of skin.

More recently in the United States a threefold increase in respiratory cancer has occurred in metal miners exposed to arsenical dust, together with other metals and radioactive products, and in South Africa a similar increase of lung cancer was found among gold miners who were exposed to arsenical pyrites. In a later survey, further cases of lung cancer were found among these miners, some of whom suffered from arsenical skin changes. Similar evidence is available from studies on smelter workers exposed to the same mixture of dusts as the miners. Another report on nickel and cobalt smelters from Germany, where the ores contained up to 50% arsenic, described 45 cases of lung cancer and skin cancer occurring in 11 years among the production workers, whilst only one lung cancer was found in the other workers. Five fatal cases were reported in 1971 among the workers in an insecticide-manufacturing plant. Within the last few years a three to fourfold excess of lung cancer has been found to occur among the workers in an insecticide plant, together with an apparent excess of cancer of the lymphatic glands. In 1975 the epidemiology branch of the US National Cancer Institute found an excess mortality for lung cancer in the general population of counties where copper, lead and zinc ores are smelted; this might be attributed to the dissemination of arsenic.

Lung cancer has occurred in other occupations, as in a taxidermist who used arsenic preservatives, and in arsenic reduction workers and spray users. It has been noted particularly in vine sprayers who both inhale arsenic in the spray and drink it in their wine. Lung cancer occurred in 12 out of 27 sprayers in one series of post-mortems, and was found in 9 out of 16 who had developed arsenical skin changes. There were descriptions of the disease in Germany and France, particularly in Beaujolais, and the evidence of a hazard appears to be very definite. 82 cases of lung cancer occurred in vineyard workers in Germany who all showed skin changes due to arsenic.

The occurrence of lung cancer among persons heavily exposed in the past may well be accepted, but the role of arsenic as an agent in the general rise of incidence of lung cancer is a more important subject. Arsenic may be inhaled from cigarettes made from tobacco treated with arsenical insecticides, and from arsenic present in the atmosphere from coal burning and other sources. American cigarettes contained an average of 12.6μg in

1933, rising to 46μg in 1950, and at the same time consumption increased threefold. Up to 10 ppm of arsenic has been reported in the general atmosphere of New York, and atmospheric pollution and cancer of the lung are well correlated. The knowledge that other metals such as nickel and chromium may have a carcinogenic or co-carcinogenic action would suggest that the presence of arsenic also in the atmosphere should be treated with possible concern.

There are, however, many strong arguments against this opinion; arsenic is no longer used as an insecticide for tobacco, and atmospheric pollution is reduced—but the incidence of lung cancer increases. Professor J. M. A. Lenihan and Dr Hamilton Smith have shown in an investigation of 1000 subjects that smokers do not have any excess of arsenic in their bodies, over that of the non-smoking population.

The organic arsenicals such as the trivalent arsphenamine and the pentavalent atoxyl do not appear to have produced a cancer hazard. Large numbers of patients have received courses of arsenic treatment without developing cancer.

Animals
Domestic and wild animals have also suffered from arsenic poisoning, both when given veterinary treatment and by pollution of pasture and water supplies. It has been estimated that approximately 200 cattle died annually in Britain from the effects of eating contaminated plants, particularly by breaking into potato fields. Poisoning has occurred from such unusual events as the accidental dropping of a large quantity of arsenical pesticides from the air, and blindness was caused in pigs by an excess of arsanilic acid in their food. The fumes in Cornwall described by Paris not only caused tumours in animals but also had the unexpected beneficial effect of killing the mosquitoes on the marshes and thus ridding the district of malaria. The reduction of bee populations is a less happy result of exposure. Birds have not been immune. Turkeys developed weakness of the legs and general trembling from excess of an arsanilic growth-promoter.

Animals have been exposed to arsenic in experimental attempts to induce cancer, but without success.

Protection
The present high standard of protection from the hazards of arsenic has been brought about by technological and economic changes and through legal measures. Arsenic has been replaced by more effective and less costly alternatives in almost all its uses. Inorganic compounds have almost disappeared from the British Pharmacopoeia and organic compounds are rarely required. Arsenic is no longer of value as a pigment. In industry it is

in increasing demand for such use as glass-making, and in agriculture it has been almost entirely replaced by modern synthetic pesticides. In Great Britain, of all the arsenicals on the list of Approved Substances, lead arsenate alone remains. Technology has removed the danger of accidental food contamination, though organic compounds may still be used as rat poisons and growth stimulators, and the use of arsenic is an economical method of tree killing in the tropics.

Protection through the law was first provided by the control of prescribing; in Brtain, after the mass poisonings from beer, the Royal Commission's recommendations on levels in certain foods were given statutory backing.

The provisions of the Factories Act contain general requirements for several procedures, and the manufacture of inorganic oxides, acids and salts of arsenic is specifically controlled by the Chemical Regulations 1922. The American Conferences of Industrial Government Hygienists have set a threshold level value (for an 8-hour working day in a 5-day week) for arsenic (0.25 mg/m^3) and arsine (0.2 mg/m^3). These levels have since been adopted by the British Government. Levels have been set for drinking water and for food by international bodies and Government Agencies throughout the world.

Summary

Arsenic has been a major health hazard to man in the past in industry, food, agriculture and medicine. This hazard has almost entirely disappeared because of the replacement of arsenic by other substances. The risks that remain are the ever-present danger of the unexpected generation of arsine, the possibility that naturally-occurring arsenic in some water supplies may be a factor in increasing the incidence of liver disease, and the possibility that arsenic in the atmosphere from industrial plants may increase the incidence of certain cancers.

FURTHER READING

Barrowcliff, D. (1971), "The Stoneleigh Abbey Poisoning Case", *Medico-Legal Journal*, Vol. 39, part 3.
Buchanan, W. D. (1962), *Toxicity of Arsenic Compounds*, Elsevier.
IARC Monograph on the Evaluation of Carcinogenic Risk of the Chemical to Man: Arsenic, Vol. 2, p. 48, International Agency for Research on Cancer, Lyon, 1973.

CHAPTER FIVE

AFLATOXINS

C. A. LINSELL

UNTIL 1960 THE MAIN INTEREST IN THE CHEMICAL METABOLITES OF FUNGI had been directed towards the development of new antibiotics. However, in that year a strange series of deaths of turkeys and trout, and the incrimination of a common cause, marked the beginning of a new biological venture. The end of the story has not yet been reached, but we have been led from an acute fatal disease in animals to the consideration of a chronic fatal disease in man—cancer.

The death of 100 000 young turkeys in the South and East of England from an apparently new disease, *turkey X disease,* was followed by reports of the sudden death of partridge and pheasant poults—as many as 5000 on a single farm. Almost immediately the death of 14 000 ducklings was reported from another farm. The fact that fatalities among ducklings were recorded from as far away as Kenya and Uganda added to the mystery. Heavy mortality among trout in commercial hatcheries in California also attracted attention, and the "disease" spread rapidly to other areas. By the end of 1960 the situation in trout hatcheries throughout the United States was a matter of national concern. Scientists and laboratories were mobilized rapidly to determine the cause. The high levels of contamination and mortality, and the coincidence of widely separated attacks, some of considerable economic importance, attracted worldwide attention, and a demand for a rapid solution. Within two years the toxin had been isolated in crystallized form, and the trout deaths had provided a clue which was to link the farmyard with the hospital—for the fish had been killed by liver cancer.

The toxic cause of turkey X disease

Poultry with the disease showed a loss of appetite, were lethargic, and died within a week. The attitude in death was characteristic—the head was drawn back, the neck arched, and legs extended backwards. Autopsies showed changes in the liver and kidneys, with acute degeneration of the liver cells, and proliferation of the bile-duct epithelial cells. Ducklings and pheasants were similarly affected. A laboratory search for pathogenic microorganisms, including viruses, was unsuccessful, and biological transmission of the disease could not be effected. There remained the possibility that the birds had been "poisoned", but intensive efforts to detect known organic and inorganic poisons and poisonous plant materials in the feed failed to disclose the presence of any toxic agent in significant amounts. It was the epidemiology, the geographical distribution of the disease, which afforded a major clue.

In England almost 80% of the deaths occured within 100 miles of London, with very few in the North, and none reported from Scotland or Wales. A preliminary survey of the outbreaks showed that all were associated with commercially prepared feeds marketed by a single London mill. A few weeks later, further cases appeared in the North which could be associated with foodstuffs manufactured by another of the company's mills. The common ingredient of the feed from these two mills was a peanut meal imported from Brazil on S.S. *Rossetti* in the autumn of 1959. The reports of deaths of pigs and cattle from unknown toxic causes in both these areas added to the economic importance of the problem, and it was not long before these too had been associated with feeds containing the Brazilian peanut meal. Further examination of this *Rossetti* meal showed that it was highly toxic to turkey poults and to ducklings, producing the symptoms of turkey X disease.

The marked susceptibility of ducklings to the toxic peanut meal formed the basis of a sensitive and rapid biological method for the detection and measurement of toxicity. Extraction procedures were elaborated to concentrate the toxic principle of the peanut meal, and purification of the toxin was carried out by chromatography. Ultra-violet light illuminated two unidentified flourescent spots, one blue and the other green. The amount of the fluorescence corresponded to the toxicity of samples when these were determined biologically in ducklings.

Although a routine chemical assay for the toxin had thus been established, its exact nature was still unknown. Speculations that the toxin might be of fungal origin were confirmed when, during the search for the presence of fragments of poisonous plants, it was noted that some peanut tissue contained fungal hyphae. These hyphae were dead, but a highly toxic sample of peanuts from Uganda subsequently produced cultures of a number of fungal species. Extracts from these cultures were examined by chromatography and one sample exhibited a fluorescent material which was identical with the toxic agent from the turkey X disease peanut meal. The toxin-producing fungus was finally identified as *Aspergillus flavus* and the toxin was given the name *a-fla-toxin* to reflect its origin.

With the advantage of hindsight, it is now possible to say that although the biological activity of these toxins was not recognized until the early 1960s, there is earlier evidence that these compounds may have been involved in previous outbreaks of liver disease in animals.

In 1952, during the investigation of a disease in pigs in the south-east United States, the most striking feature of which was acute liver disease, a fungal cause was demonstrated. Cultures of the fungi were found to be toxic and to produce liver damage and haemorrhage, but at that time attention was concentrated on the *Penicillium rubrum* toxins rather than those arising from *Aspergillus flavus*. In 1954 an "exudative hepatitis" of guinea pigs of unknown aetiology was described, the pathology of which is now seen to be remarkably similar to that induced by experimental feeding of guinea pigs with toxic *Rossetti* peanut meal. During investigations of this outbreak, the diet of the guinea pigs had been incriminated, and when the diet, containing a peanut meal, was later fed to rats, liver cancer was induced. In 1955 there was an outbreak of liver disease in kennel-reared dogs in the United States and this too was related to a commercially prepared diet. The toxic principle was not isolated, but the pattern of the liver pathology of fatal cases, necrosis and bile cell proliferation, is again similar to that produced in 1966 in dogs fed experimentally with pure aflatoxins.

A high incidence of liver cancer in hatchery-raised trout is by no means new; since the early 1930s there have been reports of fatalities from Italy, France and Japan, as well as from the United States and, although it is not possible to attribute these cancers directly to the aflatoxins, it would now appear that these compounds were the most likely cause.

There is, therefore much presumptive evidence to suggest that well before the outbreaks in 1960, foodstuffs contaminated with aflatoxins were producing the toxic symptoms and cancers which we now associate with the aflatoxins. It is of interest that during a search for more direct evidence of the association of foodstuffs with these previous outbreaks, the presence of aflatoxin was demonstrated in peanut meals which had been stored for 40 years.

The occurrence of the aflatoxins in nature

The fungi giving rise to aflatoxin are found worldwide in soil and air.

Although they cause deterioration of both cereals and seeds, they are unlikely to invade these foodstuffs to any significant extent before harvesting. First isolated from peanut meal, they have also been found to contaminate many cereals, nuts and seeds, including cassava, maize, cottonseed, rice, soya beans, wheat, sorghum and barley. Although these fungi have such a wide geographical distribution a number of factors restrict toxin production. Peanuts have been considered the substrate of choice, but there are in fact varieties of these nuts which when inoculated with toxin-producing strains of the fungi do not produce the aflatoxin. The ease with which other substrates can be infected by the toxin-producing fungi varies markedly, as does the ability of different strains of *Aspergillus flavus* to produce toxins.

The micro-climate necessary for the production of aflatoxin is critical and is chiefly related to humidity, both that surrounding and within the substrate. A relative humidity of more than 70% is usually necessary for the growth of fungi on stored food, and a greater level of humidity (over 85%) with a moisture content of 30% in the substrate is probably necessary before the fungus can produce toxin. When the relative humidity falls to 50% and the moisture content to 15%, little toxin is produced. Rapid harvesting and drying of a crop before storage will greatly reduce the aflatoxin contamination of the stored cereal or nuts. As many crops are sun-dried, local climatic conditions will have great influence on the subsequent level of aflatoxin contamination. The conditions of humidity and moisture content will also affect the ratio of the various aflatoxins in the stored crops. It has been shown that damage to cereal or nuts, either mechanical or by insects, renders them more likely to be contaminated by the fungus and aflatoxin. This usually takes place during storage after harvesting, but aflatoxin has been found on growing maize when this was badly damaged by weevils.

Wherever peanuts have been examined, a proportion has been shown to be contaminated, but the pattern of contamination of the other substrates is more variable. The levels of contamination of harvested rice are usually low, but high levels have been reported in parboiled rice, or rice which has been put aside after a meal. Certain kinds of meat, when inoculated with aflatoxin-producing strains of the fungi, can sustain aflatoxin, but no aflatoxin has been demonstrated in commercially available meat or meat products. Although it is relatively simple to produce aflatoxin on soya beans in the laboratory, they are considered to be relatively unfavourable substrate for natural aflatoxin production.

Brazil nuts are an interesting potential source of contamination, particularly as the nut may lie buried for some time before harvesting. The wooden-coated Brazil nuts grow as segments within an outer coat, the whole resembling a cannon-ball weighing a kilogram or more. These fall when ripe from the tall Brazil nut tree and bury themselves in the soil around the trees. No one, rather naturally, will approach these trees during this season, and the nuts within the parent shell are exposed for some weeks to fungal contamination in the hot humid soil. Aflatoxin has been found in other kinds of nuts—pecan, almond, pistachio and walnuts—and it is difficult to envisage how sampling and inspection can ensure that in-shell nuts of any kind can be declared free of mycotoxins.

The contamination of cottonseed and cottonseed cake, a common animal feed, presents another major problem. The contamination of spices and condiments, peppers and chilis, has been reported from Asia. Although aflatoxin has been demonstrated in the milk of dairy cattle, such evidence as we have shows that this is unlikely to be a hazard to man consuming either fluid or dried milks.

The chemical composition and properties of the aflatoxins

The aflatoxins, a group of bis furano-isocoumarin metabolities synthesized both by *Aspergillus flavus* and *Aspergillus parasiticus*, are designated B_1, B_2, G_1 and G_2. When a sample is analyzed by thin-layer chromatography, the aflatoxins separate into the individual components in the order given above. The first two fluoresce blue when viewed by ultra-violet light, and the G aflatoxins fluoresce green. These properties form the basis of their chemical analysis. In natural contaminations B_1 is found in the greatest concentration and B_2 and G_2 in the smallest amounts. They do not necessarily occur in the same sample, and their relative concentration to each other depends on the strain of the fungus and the substrate on which it is grown.

aflatoxin B_1 aflatoxin G_1

The empirical formula of aflatoxin B_1 is $C_{17}H_{12}O_6$, and the chemical structure shown has been confirmed by laboratory synthesis. The formula for aflatoxin G_1 is $C_{17}H_{12}O_7$ and the aflatoxin B_2 and G_2 are dehydro-derivatives of the parent compounds. The excellent review by Roberts (1974) covers the chemistry of the aflatoxins and the sterigmatocystins. The chemistry of sterigmatocystin is usually included when considering the aflatoxins, as it is a precursor in the biosynthesis of aflatoxin which was isolated and characterized from the mycelium of *Aspergillus versicolor* many years before the discovery of aflatoxin. Many of the structural studies on aflatoxin B_1 were based on the isolation and characterization of sterigmatocystin.

The M derivatives of aflatoxin are the products of metabolism in animals, and are usually found in milk and urine. The designation M was derived from the first isolation of this metabolite in the milk of cows. Aflatoxin P_1 is one of the principal urinary metabolites of aflatoxin B_1 in rhesus monkeys, and its importance is related to the use of these animals in toxicity and carcinogenicity experiments.

One of the important properties of the aflatoxins is their stability under normal conditions of handling and food processing. They are particularly stable when absorbed onto the starch or protein surfaces of seeds or cereals, which is the usual way that crops are contaminated by these toxins. There is little or no transformation of the aflatoxins synthesized in or on seeds or cereals to other forms of the molecule.

The heat stability of the aflatoxins on substrates is remarkable. Negligible destruction of aflatoxin on cottonseed is reported at temperatures between 60°C and 80°C; at 100°C there is a marked reduction when this temperature is used for two hours with a moisture content of 20%. Degradation of 80% of B_1 and 60% of G_1 follows roasting of cottonseed meal at a temperature of 150°C. Autoclaving moistened peanut meal for four hours markedly reduces the toxic level but does not eliminate it completely.

The pure aflatoxins on the other hand are sensitive to fluorescent light, ultra-violet radiation, heat and chemicals, and degrade so quickly that precautions have to be taken in analytical procedures to prevent breakdown.

The presence of a lactone in the aflatoxin molecule makes it susceptible to alkaline hydrolysis; any food process using alkaline treatment will detoxify the compound.

The metabolism of aflatoxins

In addition to the academic interest in the metabolism of the aflatoxins, there is the practical importance of the extent to which animal products become contaminated when animals are fed rations containing the aflatoxins. Apart from the oral route, aflatoxin can be absorbed subcutaneously or by topical application, as is shown by the production of tumours at these sites. Experiments using the oral route are important however, as these have the greatest validity when considering the public-health significance of contamination.

It would appear that only small proportions of the compounds are excreted without conversion to derivative forms. Experiments with rats and mice show that 80–90% of a single dose of the compound is excreted within 24 hours. The biliary and faecal routes are the principal pathways, accounting for more than 65% of the total amount excreted. The remainder is excreted in the urine. The main target organ for toxicity, and indeed carcinogenicity, is the liver which may concentrate the toxin/carcinogen. Much of our information on the metabolism of aflatoxin has been derived from experiments in rats, in which a single dose of aflatoxin, although rapidly excreted, gives rise to histologic damage and can induce cancer, even after a considerable time-lag. Biochemical alterations in the liver persist far beyond the time that the presence of the compound can be detected in the tissues by currently available methods.

There is evidence that aflatoxin B_1 causes traumatic alterations in nucleic acid and protein synthesis in the liver, as a result of the interaction of the toxic material with DNA in such a way as to interfere with its transcription; this would cause an impaired synthesis of DNA and DNA-dependent RNA. The interaction of aflatoxin with DNA does not appear to be a direct one, but rather a process of activation through an epoxide metabolic pathway. It has been suggested that the marked difference in sensitivity between and within animal species to aflatoxin may be related to the balance of metabolism between activation and detoxification. This and other evidence suggests that the ultimate carcinogenic form is an activated aflatoxin. This would be in line with many carcinogens which are not reactive until converted to an intermediate compound.

As many animal feeds contain cereals which could be contaminated by the aflatoxins, the question arises whether the toxic/carcinogenic material could be found in their body tissues. Such evidence as we have suggests that the consumption of animal tissues does not seem to be a hazard to man in regions where potentially toxic *Aspergillus flavus* strains contaminate foodstuffs of animals. While animal tissues may contain no aflatoxin, the consumption of cereal products, seeds, nuts and derivatives of these materials may well represent a definite hazard to man.

The acute toxicity of aflatoxin

As will be seen from Table 5.1 there is wide variation in the LD_{50}—a dose capable of killing 50% of the animals exposed to a particular dose of a toxin—throughout the range of animal species tested with single doses of aflatoxin. Some species show a wide range and it is among these that different strains demonstrate varying sensitivity or resistance to the aflatoxins. Of all laboratory animals, rats have been most extensively studied, and there are marked differences in susceptibility to the acute toxicity between the various strains. These differences make it difficult to assess many of the studies on toxicity, but in general the pathological lesions in both sexes and in different strains are similar, namely a periportal necrosis and biliary proliferation developing over three days, in addition to kidney and adrenal damage. Metabolic studies have shown that rabbits, ducklings and guinea pigs metabolize an LD_{50}, if they survive.

Table 5.1   Single LD_{50} values and metabolism of aflatoxin

Species	Single LD_{50} (mg/kg body weight)	Differential metabolism	Tumour induction
Rabbit	0.3 – 0.5	fast	−
Duckling	0.35 – 0.56	fast	+
Cat	0.55	−	−
Pig	0.62	intermediate	−
Rainbow Trout	0.81	−	+
Dog	1.0	−	−
Guinea Pig	1.4 – 2.0	fast	−
Sheep	2.0	intermediate	−
Monkey	2.2	−	+
Chick	6.5 – 16.5	intermediate	−
Mouse	9.0	intermediate	+
Hamster	10.2	−	−
Rat	5.5 – 17.9	slow	+

LD_{50} will depend on methods of administration, age, sex and solvent carrier.

The liver pathology of the acute effect of the aflatoxins differs from that observed with exposure to other known carcinogens, such as carbon tetrachloride or the nitrosamines, and the distribution and type of lesions within the liver vary with the different aflatoxins. In animal experiments the aflatoxins do not appear to be associated with development of cirrhosis of the liver—an important point as we shall see later. Lesions are not confined to the liver; the tubular epithelium of the cortex of the kidney and the adrenals also show evidence of necrosis. Monkeys and primates are moderately sensitive to the acute effects of aflatoxin poisoning, but our main interest is where man fits into this table.

Acute toxicity in man
Information is scanty, as those areas of the world in which contamination is most frequent are those where medical services are less developed, and therefore many cases may pass unnoticed. However, we have well-documented cases from Africa and Asia of liver disease similar to that found in animal experiments with aflatoxin. Most of the patients were children who had either accidentally consumed home-prepared foods which were later found to be contaminated by high levels of aflatoxin, or

who were fed on protein supplements containing contaminated cereals, usually peanuts, during the treatment of kwashiorkor or other deficiency diseases. In the first situation, as those doses were so high, it is difficult to explain the lack of symptoms in adults who fed on the same food as the children who died. This resistance may be based on the age-related differences in the metabolism of aflatoxin in man.

The experience in Africa and Asia of aflatoxin-contaminated peanuts being inadvertently incorporated as a source of protein into supplementary food for malnourished children has created a problem for nutritionists and food manufacturers. Peanut meal is a good source of protein to supplement special diets for the malnourished in tropical countries but, as these nuts can be contaminated, their use has been controversial. As available proteins are often rare in the tropics, the neglect of an important source may be difficult to justify, and to balance the dangers of infant malnutrition against the potential dangers of a cancer in adult life is a difficult issue. The modern methods of detection and control of aflatoxin contamination now available have gone a long way to solving this problem.

Autopsy studies of a number of children in Thailand who had died suddenly after ingestion of food heavily contaminated by aflatoxin emphasized the similarity of the liver pathology to a syndrome which had been reported by Reye as early as 1900. Cases of this syndrome had been described from a number of widely separated countries, and there had always been the suggestion that the background of this disease might be associated with fungal poisoning. The scientists in Thailand were also carrying out aflatoxin feeding studies on monkeys, and they were struck by the similarity between the pathology of the liver in these animals and that of the children with the symptoms of Reye's syndrome; more direct evidence was obtained when the food and the tissues of the children were examined, and aflatoxin detected.

A recent report from India of an outbreak of hepatitis associated with aflatoxin which resulted in the death of over 100 patients indicates that these incidents may not be as rare as we suppose. However, they are probably all associated with some grave dislocation of food supplies, which results in the consumption of food which in the normal course would be rejected. Most cereals and nuts heavily contaminated by aflatoxin will be unacceptable to the normal palate, and it is noteworthy that grossly contaminated samples of the original *Rossetti* meal were unacceptable to laboratory animals unless the diets were flavoured with molasses.

There are other examples of widespread disease associated with fungal contamination of food. The "yellow rice disease" in Japan, which caused many deaths, was due to the invasion of the rice by a number of *Penicillium* moulds. This rice had been imported from South East Asia to make up for the deficit in Japanese production during the war. It had been exposed to poor storage conditions, and moulds flourished. The even more severe loss of life reported from Russia associated with the ingestion of grain contaminated by *Fusarium* fung. was also due to exigencies of war conditions, following widespread conscription of farm workers for the battle front. Grain, which should have been harvested in the autumn was left in the fields under the snow during the winter, and was salvaged in the spring. Acute mycotoxicosis, including that from aflatoxin, should therefore always be thought of in these desperate situations. In Thailand, the outbreak of acute aflatoxin poisoning was associated with the practice of eating rice which had been prepared some days previously. Such food practices, and the ever-present threat of starvation for the rural farmers of Asia and Africa which lead to the eating of food which has been poorly stored and which under more fortunate circumstances would not be acceptable, may present grave problems.

The carcinogenicity of the aflatoxins

Although the most common cancer produced by the aflatoxins in experimental animals is that of the liver, other sites in the digestive tract and kidney have been recorded. Variations of susceptibility to the carcinogenicity between species and between strains within species have been recorded. Many carcinogenicity studies on the aflatoxins have been carried out on the rat. Although the early experiments with contaminated commercial feeds demonstrated the dangerous carcinogenic properties of the aflatoxin, it was only when the purified aflatoxins were available that it was realized that the toxin was probably the most potent liver carcinogen known. When purified aflatoxin B_1 was used in in-bred Fischer rats, levels as low as 0.015 parts per million of the diet produced tumours and a low-dosage régime for a limited period produced tumours months later. In one study it was shown that a single dose of aflatoxin could induce liver cancer after many months of subsequent normal diet. A significant feature of hepatic carcinogenesis with the aflatoxins is that cirrhosis of the liver is not a necessary precursor of tumour induction. However, it has been demonstrated that when cirrhosis is produced by other means, subsequent aflatoxin exposure increases the carcinogenicity of the aflatoxins. It was some years before tumours in monkeys and primates were demonstrated, the first unequivocal experiment requiring an exposure of over six years.

Recent studies in tree shrews (which are regarded as primitive primates) fed varying doses of aflatoxin B_1, produced a 100% incidence of tumours in females and a 50% incidence of tumours in males. It is concluded by the authors of this experiment that the individual variation in cancer response to the differing amounts of aflatoxin is very marked, and that an extremely cautious approach should be maintained when proposing permissible or safe levels of contamination of carcinogens in foodstuffs. Whereas the studies in rats and trout have demonstrated that the degree of liver damage parallels the accumulated dose of the toxin, the studies in tree shrews and monkeys may demonstrate that there is a considerable range in the ability of the liver to handle aflatoxin within an out-bred strain of animals. As the induction of cancer by aflatoxin has proceeded so far up the phylogenetic tree as the non-human primates, and as this toxin is freely available in foodstuffs, we will now consider the possibility of it being associated with cancer in man.

The geographic distribution of human liver cancer

The incidence and the pattern of cancer types varies markedly from one geographic region to another, and none more so than liver cancer. It has been estimated that the rate in Mozambique in Southern Africa, where the highest incidence in the world has been recorded, is 500 times that of New York City or Birmingham. Liver cancer is most common in the south of Africa (becoming rarer as one goes north to the Sahara) and in the Far East. The high incidence in sub-Saharan Africa appears to be independent of ethnic background of the indigenous Africans and was present before Western influences made themselves felt in the continent. Studies have also shown that people of Africa and Asia who emigrate to countries of low incidence tend to acquire the pattern of their adopted country, whereas Europeans who migrate from areas of low incidence to those of higher incidence, such as Africa, retain their low rate. This would suggest that we must study environmental factors which are associated with the rural non-industrial basic ecology of those countries where the cancer is particularly common.

Liver cancer in Africa and Asia

Studies of selected foodstuffs from such widely separated areas as South Africa, Uganda, Thailand and the Philippines showed a wide range of foodstuffs to be contaminated by aflatoxin. Preliminary studies of the association of these contaminated foods with the incidence of liver cancer have led to many conjectures, based on the random sampling of stored staple food, without excluding the possibility of subjective bias. There have been few field epidemiological studies to assess the correlation between the ingestion of potential carcinogens and the incidence of a cancer, and a number of factors make the interpretation of such studies difficult. First the duration of the period between the exposure to a carcinogen and the development of a cancer is unknown. Secondly, the assessment of average exposure of a community does not necessarily reflect the distribution and level of individual exposure. Cancer registration is also difficult in the areas in which high levels of this liver cancer have been reported. Possibly the most difficult aspect to assess is that of other factors, such as the effect of other mycotoxins, the effect of other items of the diet, and the possibility of predisposing liver disease. Cooking procedures and food habits may also play a part, and in Africa routine sorting and rejecting of contaminated cereals by housewives has been shown to reduce markedly the available aflatoxin in food ready for ingestion.

To minimize the complications listed above, a series of studies was carried out in Africa and Asia to assess the relationship for food ready for ingestion in a homogenous rural population using a simple diet which had changed little over the preceding decades. Of necessity, the methods of cancer registration varied between the studies, but liaison was maintained between the different field teams. Analysis for another mycotoxin, sterigmatocystin, which had been shown to be carcinogenic, was included in one study, but no sterigmatocystin was detected in any of the diets sampled. Two studies in Africa, in Kenya and Swaziland, demonstrated a relationship between the ingestion of small doses of aflatoxin and cancer incidence in relatively small populations of approximately 400 000. Households were randomly selected and food ready for ingestion was collected for analysis for the aflatoxins. Care was exercised to ensure that special food was not prepared and that the food analyzed was in fact about to be eaten by the subject in the study areas. The studies extended over 18 months to obtain samples of food at all seasons of the year and under all prevailing conditions of storage. A detailed examination was made of storage facilities, marketing, nutritional and sociological data. Registration of all cases of cancer extended over five years, to eliminate as much as possible reporting variations and to assess time trends.

Similar population-based studies in Thailand supported the association between levels of aflatoxin ingestion and cancer rates. Investigations in Mozambique, in a population having the highest incidence of liver cancer in the world, have also shown that even at this level of cancer incidence there is a statistically significant parallel with aflatoxin ingestion.

Liver cancer in Europe and North America

There is evidence that liver cell cancer is increasing in Europe and North America, and that this is not merely a consequence of improved diagnostic facilities. The clinical diagnosis of liver cancer in Africa does not present problems for the physician experienced in African medicine, and a survey has shown that 80% of such diagnoses proved to be correct on subsequent autopsy. However, in Europe and North America, diagnosis is not easy, as tumours of the gastrointestinal tract, lung, etc., metastasize readily to the liver. These tumours are much rarer in Africans, so an enlargement of the liver, rapid wasting and usually death within months, is all too often the whole sad story.

The recent development of a rapid specific test for liver cancer has assisted diagnosis in areas of high incidence and has proved of great value in the differential diagnosis of liver malignancies elsewhere. The test is based on the reappearance in the serum of a foetal protein—alphafoetoprotein—in liver disease and at high levels in liver cancer.

Although the discovery and characterization of the aflatoxins were made in the Western world, it is very unlikely that they played any part in the induction of liver cancer in Europe and North America. Monitoring of foodstuffs for aflatoxin, and rejection of contaminated supplies, are standard practices in most countries of this region, and food supplies and dietary habits are different from those of the rural populations where epidemiological studies linked the cancer with aflatoxin.

It is difficult to judge a situation with so many factors, but it is generally agreed that liver cancer occurring in the technically developed nations has a different range of factors from those occurring in the rural populations of developing countries. There may, however, be factors common to both, and we must therefore now consider one—cirrhosis.

Cirrhosis and liver cancer

Liver cancer is associated with cirrhosis in two ways. At autopsy or biopsy of liver cancer patients, concomitant cirrhosis is found in the majority of cases. This is true in both the Tropics and the Western World, although there is evidence that in areas of high incidence in Southern Africa cancer more frequently occurs without any signs of cirrhosis. Secondly, unsuspected cancer is frequently demonstrated at autopsy of patients with long-standing cirrhosis. It would appear that patients with cirrhosis are at high risk for liver cancer, and that it is wise to suspect that any sudden deterioration in the clinical condition of a cirrhotic may be associated with the onset of cancer in the already damaged liver. The cirrhosis of the Western World is associated with alcoholism, whereas the cirrhosis of the Tropics is usually considered to be the result of a previous viral infection. The morphological type, whatever the aetiology, may be an important factor. Their classification has been the subject of controversy for many years, but it is clear that the large nodular cirrhosis found commonly in the Tropics, rather than the small nodular type associated with alcoholism, is more likely even in Europe to be associated with cancer wherever this occurs. A change from the small nodular to the large nodular cirrhosis has been found to occur in Western cirrhotics, and this has been linked with reformed alcoholics and with improved treatment of cirrhosis and a longer

L

survival. It is now suggested that this is probably the most important factor in the recent increase of liver cancer in the Western World.

Liver cancer in the Tropics and aflatoxin

It is now known that the aflatoxins are present in food in the Tropics, that they are eaten, and that in some areas the level of the amounts ingested parallels liver cancer rates. Whether there are areas where aflatoxin contamination is common, but where liver cancer incidence does not follow this pattern, is not known. Acute aflatoxicosis has been reported from India, and crops are known to be contaminated. Information on liver cancer incidence in the sub-continent is limited, but it certainly does not indicate the high levels of Africa or other areas in the East. Food habits or storage conditions may intervene in the cycle between contamination and ingestion, or other factors necessary for the carcinogenic action of aflatoxin may not be present.

An attempt to remove aflatoxin from the diet of a population with a current high rate of liver cancer would appear to be the most practical way of testing the association. It would not be easy to mount a study in rural populations, such as those used so far for assessing the association, without interfering extensively with the life of the community. Among the many other problems which such an approach evokes are the very considerable period before an intervention trial can be evaluated and the expense of such a project. However, even if the toxin proved not to be important in the aetiology of human cancer, a dangerous food contaminant would have been removed, and interesting methodologies evolved for the control of environmental carcinogens.

It would be necessary to evaluate any other factors, and we now have the possibility of monitoring at least one of them. The recognition of the viral agent responsible for one type of hepatitis, the hepatitis B antigen, represents a major advance. This type of hepatitis is more frequent in the Tropics than in the Western World, and may play a more important role in liver disease there. Evidence of the viral infection is found in a high proportion of liver-cell cancer patients wherever the tumour occurs, and in cirrhosis in those countries there the cancer is frequent. It is perhaps unlikely that the virus is directly oncogenic, but the pathology which it precipitates could well be a cofactor in the aetiology of the cancer. As animal studies have shown that the aflatoxins are not good cirrhogenic agents, they may not be so in man, but we have already noted that the carcinogenic activities of the aflatoxins are enhanced when cirrhotic animals are exposed to the toxins. Whether the sequence of the induction of cancer in man is the development of a cirrhosis following an attack of hepatitis with marked hyperplasia, which is such a feature of this condition, or the persistence of a chronic hepatitis with some evidence of repair and hyperplasia followed by a further insult from a mycotoxin, such as aflatoxin, or whether the mycotoxin insult precedes the viral insult, we just do not know. However, rapid advance in our knowledge of the part viruses may play in liver disease in the Tropics can confidently be expected and, as these may well lead to the production of a protective vaccine, control of this other suspect factor associated with liver cancer may prove possible.

Conclusion

The evidence that aflatoxin causes cancer in animals, including non-human primates, is clear. In rats—the animals most frequently used for testing food contaminants—very small doses of aflatoxin produce cancer, and it is thought to be the most potent liver carcinogen known to the experimentalist. It is available on foodstuffs in the Third World, where liver cancer is more common. As far as is known, it is eaten regularly by many and has been consumed in large apparently-fatal doses in rural communities. It is highly toxic.

With the control of imports and the monitoring of suspect cereals, seeds and nuts, there does not appear to be a danger to the health of the technically developed countries. The dangers of using meat or milk, either dried or liquid, from animals which may be exposed to aflatoxin-contaminated feeds are, it appears, theoretical rather than real. It is likely that the control of this hazard to imports into the developed countries will become more complex as stricter monitoring follows further knowledge of the potential dangers to man. Thus, as exports from some of the developing countries are rejected, the countries will be faced with additional economic hardship as well as remaining a site of maximum exposure. The developing countries most affected will be those exporting peanut meals, oils, and other products used mainly for cattle feed.

There are large gaps in our knowledge of the geographic distribution of liver cancer but if, as is suspected, it is frequent in the large populations of the East, Indonesia and China, as well as in Africa, it may well represent one of the most frequent cancers in man. It would be rash to consider the relationship between the cancer and the toxins as causal at this time, but the epidemiological evidence and the background of laboratory studies are as strong as for any chemical carcinogen, and these will undoubtedly weigh heavily with those countries where it is known to be present in staple foods.

There are many other factors which must be considered, notably the virus diseases of the liver, and the very real possibility of a protective vaccine for at least one type of hepatitis is a welcome prospect. If the aflatoxins play a role in the induction of human liver cancer, it is the rural farmers of the Third World and their families who are at the greatest risk.

The solution to the problem will require the resources of the agricultural planners—particularly the food-storage experts, the nutritionists and sociologists. Long-established food-storage practices, and even dietary habits, may have to be modified.

With the identification of possible causal factors—one chemical and the other viral—we can be encouraged in that prevention, difficult though it may be, does appear a practical possibility. There are not many cancers for which such hope can be offered.

FURTHER READING

Cameron, H. M., Linsell, C. A., and Warwick, G. P., Eds., *Liver Cell Cancer, 1976,* Elsevier Scientific Publishing Company, Amsterdam, New York, Oxford, pp. 292.

Editorial: "Aflatoxin and Primary Liver Cancer," *South African Medical Journal,* **48,** No. 60, 11 December 1974, pp. 2495–2496.

Editorial: "Cancer and Food," *Lancet,* II, No. 7838, 17 November 1973, pp. 1133–1134.

Goldblatt, L. A. Ed. (1969) *Aflatoxim—Scientific Background, Control and Implications,* Academic Press, New York, pp. 472.

Linsell, C. A. and Peers, F. G. (1972), "The Aflatoxins and Human Liver Cancer", in *Recent Results in Cancer Research, 39.* Ed., Grundmann, E. and Tulinius, H., Heinemann Medical, Berlin, Heidelberg and New York, pp. 125–129.

Purchase, I. F. H. (1971), *Symposium on Mycotoxins in Human Health,* Macmillan Press, London, pp. 297.

Purchase, I. F. H. Ed., (1974), *Mycotoxins,* Elsevier Publishing Co., Amsterdam, pp. 443.

Roberts, J. C., (1974), "Aflatoxins and Sterigmatocystins", in *Progess in the Chemistry of Organic Natural Products,* Eds., Herz, W., Grisebach, H., and Kirby, G. W., Springer Verlag, Vienna and New York.

CHAPTER SIX

ASBESTOS

M. L. NEWHOUSE

ASBESTOS IS A WIDELY USED AND VALUABLE MATERIAL, WITH very many applications in the modern world, yet it constitutes a major health hazard to workers in industry, and is a potential threat to certain communities. The mineral consists of bundles of fibre found in natural ore (figure 6.1).When separated, they can be woven into textiles; when mixed with other materials, they add not only their fire-proofing qualities but also tensile strength and flexibility. Asbestos when dry easily separates and forms dust, which consists of fibres varying in size from several inches in length to fibrils of microscopic size. It is these fibrils which are a risk to health. They may remain suspended in air, can be inhaled into the lungs, and can contaminate the atmosphere at a considerable distance from the area where the material is being handled.

Structure and uses

Asbestos is the only naturally occurring fibrous mineral. The different types are shown in the diagram.

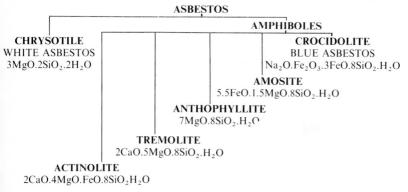

ASBESTOS

AMPHIBOLES

CHRYSOTILE
WHITE ASBESTOS
$3MgO.2SiO_2.2H_2O$

CROCIDOLITE
BLUE ASBESTOS
$Na_2O.Fe_2O_3.3FeO.8SiO_2.H_2O$

AMOSITE
$5.5FeO.1.5MgO.8SiO_2.H_2O$

ANTHOPHYLLITE
$7MgO.8SiO_2.H_2O$

TREMOLITE
$2CaO.5MgO.8SiO_2.H_2O$

ACTINOLITE
$2CaO.4MgO.FeO.8SiO_2H_2O$

Chrysotile asbestos, which has white fibres, occurs in serpentine ore. The importance amphibole asbestoses are crocidolite (blue), amosite (light brown), and anthophyllite (white). Tremolite and actinolite are also shown in the diagram; they are talc-like and used to a much lesser extent in industry. All forms of asbestos are silicates; the crocidolite and amosite molecules also contain iron. The different types of asbestos have different properties, leading to different uses in industry.

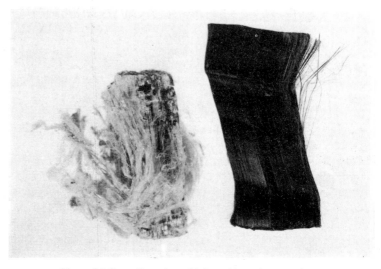

Figure 6.1 Crysotile and crocidolite asbestos in natural ore

Chrysotile is mined extensively in the Quebec Province of Canada, in the Soviet Union and in South Africa and Rhodesia. There are also small mines in Europe, Cyprus and South America. It is the most widely used type of asbestos and forms 90% of world output. Its resistance to heat is high; and because of its great tensile strength and often silky texture, it is suitable for weaving heat-resistant textiles for fire-proof curtains, garments and gloves. Its most extensive use, however, is in asbestos cement for roof sheeting and other building materials and for asbestos cement pipes. It is also used in the electrical insulation industry and, because of its friction-resistant qualities, in brake linings and clutch facings.

Crocidolite is mined both in Cape Province and in the Transvaal in South Africa. There was a mine in Western Australia, but this has now closed. The particular properties which make it commercially desirable are its high resistance to acid and its high tensile strength. It has had many uses, particularly for heat-resistant insulation and for acid-resistant filters and packings. Amosite is also mined in South Africa, and has similar properties and uses to crocidolite. Anthophyllite is mined in Finland, but the production of the mine is comparatively small and little is exported.

The fire-resistant properties of asbestos have been known since antiquity. It was used by the Romans for lamp wicks and grave wrappings, and has been found as a filler on pottery dating from the fourth century BC. The commercial use of asbestos, however, is comparatively recent. The mining industry started in Canada in the 1880s and a little later in South Africa. The first amosite mine was not developed till the 1920s. The first English asbestos factory, opened towards the end of the nineteenth century, was in Lancashire and used chrysotile asbestos. A London factory opened in 1906 chiefly using, at that time, South African crocidolite. The industry has expanded and diversified very considerably. The total world output of asbestos is now more than 4,000,000 tons and most countries have an asbestos industry. Asbestos is used in the shipbuilding industry both for fire-proofing and lagging; the big expansion of shipbuilding during the second world war was responsible for diseases recognized only in the last 10 years.

Asbestos mines may be either underground (some South African mines have reached a depth of between two and three thousand feet) or opencast, as in the eastern townships of Quebec. After mining, the ore is milled to separate the fibre—often a dusty process. When separated, it is bagged and transported to the manufacturer. Here is has to be prepared for use, either for spinning or weaving into textiles and yarns, or to make a wide variety of other products. Asbestos is dangerous only in the form of dust, but dust can be generated at any point where the asbestos is dry, so that workers in many different jobs and industries may be at risk, for example, in mining and milling, in transport, in loading and unloading at docks and in the factories. Asbestos products are very widely used in the construction industry, and workers there also are frequently exposed.

Biological effects

Asbestos corns

Asbestos fibres may penetrate the skin of the hands of workers, causing the so-called asbestos corns. The sharp fibres may cause the same type of reaction as any splinter, but the fibre is easily removed and the corn (which is of little significance) subsides.

Asbestos bodies

Long asbestos fibres, like any other dust particles over a certain size, will not pass through the nose and descend through the trachea into the lungs; the important fibres from the health point of view are the very fine fibres which are not visible to the naked eye. These fibres can pass with inhaled air right down into the small air passages and enter the alveoli or air spaces of the lungs. Once in the tissues of the lung, many fibres are not removed by natural processes and remain there for the rest of the individual's life. Some fibres become coated with an iron-containing protein known as *feratin* and form the asbestos or ferruginous body. These bodies (figure 6.2) are easily

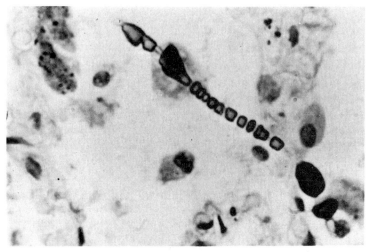

Figure 6.2 Segmented asbestos body in lung tissue (× 800)

identified under the microscope; they have characteristic dumb-bell shapes are often segmented, and are golden brown in colour. They may form in, or migrate into, the small air passages, and be coughed up with the sputum. They are an indication, not of disease but of asbestos exposure; the finding of asbestos bodies either in the lung at post-mortem, or in operation specimens, or in a living person's sputum, is an indication that at some time during life he has been exposed to asbestos dust. Using modern techniques of electron microscopy, the asbestos fibre within the asbestos body can be identified and electron probes can identify the particular type of fibre. Electron microscopy can also identify asbestos fibres or fibrils lying uncoated in lung tissues.

Asbestosis

A very important effect on the lung is the reaction of the pulmonary tissue to these silicates; they initiate a fibrogenic reaction in the lung tissue. This means that fibrous tissue is generated around the alveoli of the lungs and, instead of the one-cell-thick lining of the alveolus which permits the interchange of oxygen and carbon dioxide from the blood in the capillaries to the air of the alveolus, there is a thickened membrane, and exchange of gases is impeded. This fibrosis of the lung is known as *asbestosis*. It affects first the lower part of the lungs, but eventually may affect all areas. Severely affected persons develop diminished lung volumes, and become progressively more breathless on effort. Exchange of oxygen may be so diminished that cyanosis develops: the lips and cheeks turn blue, coughing develops, and eventually those affected become pulmonary cripples capable of little physical effort. In the end they may die of heart failure, being unable to obtain sufficient oxygen through the lungs. In some cases the disease is arrested spontaneously, though it may progress even after exposure has ceased. Although asbestosis is not the inevitable result of asbestos exposure, its development is certainly related to the severity and length of exposure.

A long-term study of the incidence of asbestosis has been made in a Lancashire factory. In 1930, at a time when the health risk was not appreciated, 32% of workers with exposure of not more than 15 years were affected. Among those with very long exposure (20 years or more), 80% developed asbestosis; but by 1967, when dust control had become effective, the incidence had fallen to less than 1% for those with the shorter period of exposure, and to 2% for those with very long exposure.

Individual susceptibility also has an important role. It is probably governed by immunological response to the inhaled particles.

Asbestos plaques

A further effect in the lung tissues is the formation of asbestos plaques—areas of thickening which develop in the pleura. (The pleura is a thin membrane which lines the chest cavity and is reflected over the lungs.) These pleural plaques are thought to be a reaction of the pleura to asbestos fibres which reach the periphery of the lung and penetrate to the membrane. Hyaline tissue is formed, and later calcium is deposited. Both the thickened areas and the calcified plaques may be visible on radiological examination. They appear to have no deleterious effects on health, and when observed on chest radiography are another marker of asbestos exposure. There is no evidence to suggest that these are the sites of origin of the cancers associated with asbestos exposure which will now be discussed.

Asbestos-related cancers

The two most important cancers are cancer of the bronchus, usually referred to as *lung cancer,* and mesothelioma of the pleura and the peritoneum. (The peritoneum is a membrane of similar structure to the pleura, which lines the abdominal cavity and the abdominal organs.) Asbestos exposure may also be associated with an increased incidence of cancers of the stomach and bowel, and of the larynx.

Cancer of the lung

In the early days of the industry, i.e. up to the second world war and continuing up to the 1950s, workers who contracted asbestosis frequently died of tuberculosis—often as comparatively young men or women. It must be remembered that tuberculosis was very widespread in the early part of the century but, since the post-war development of chemotherapy, this disease has been well controlled. Parallel with the decrease in the incidence of tuberculosis in the general population, has been the increased incidence of cancer of the lung, generally attributed to the effects of smoking. During the 1930s, cancer of the lung was not a very common condition, but case reports of deaths from cancer of the lung in asbestos workers began to appear in the medical literature from 1933 onwards, and the comparative frequency of the condition among asbestos workers was a subject for comment in reports of the Medical Inspector of Factories.

Interest increased and a definitive study was made in Lancashire by Doll; he showed that among workers heavily exposed to asbestos dust for 20 years or more, the risk of dying of lung cancer was ten times greater than the risk in the general population. Since that time there have been many studies of groups of workers in factories, in mines, among laggers (known in the United States as insulation workers), and in docks and shipyards. In all studies, cancer of the lung is found to be a very high risk.

In 1968, a further important observation was made by Dr Irving Selikoff of the Mount Sinai Hospital of New York. Studying the mortality of workers in one branch of an asbestos insulation union, he found that the risk of death from lung cancer was 90 times greater in asbestos workers who smoked than in non-smokers not exposed to asbestos; his original figures are shown in Table 6.1. He confirmed this observation by studies of much larger groups of workers.

Table 6.1 Mortality of asbestos workers by smoking habits (data from Selikoff, I. J., Hammond, E. C. and Churg, J., 1968, *JAMA*, **204**, 106)

Smoking habits	No. of men	Observed deaths	Expected deaths
Never smoked regularly	48	0	0.05
History of pipe or cigar smoking only	39	0	0.13
History of regular cigarette smoking	283	24	3.16

In London, the author and her colleagues were following the mortality of past workers at an East End asbestos factory. With the active co-operation of the Department of Health and Social Security, the smoking habits of surviving workers were ascertained. In this study it was shown that exposure to both smoking and asbestos appeared to produce a multiplicative risk; i.e. if the risk of developing lung cancer due to asbestos exposure was given a value of 5, and the risk due to smoking a value of 10, the combined risk for exposure to both was 50 rather than 15. In other words, the carcinogens of asbestos and of cigarette smoke were acting synergistically rather than independently.

In the late 1960s Macdonald of McGill University in Montreal started a comprehensive study of the chrysotile asbestos mining industry in Quebec. This was centred on two mining towns, one Thetford and the other actually named Asbestos. Studies were made of persons still at work, and the causes of death of all those who had ever worked in the mines were also determined. It was possible to obtain records of the measurements of dust concentrations in air in mines and mills, and the number of years each man had worked was known. From these two sets of data it was possible to work out an index of dust exposure which Macdonald calls "million particle per cubic foot-years" (m.p.c. f-years) (Table 6.2). The average death rate for lung cancer increases from 10.3 for 10 m.p.c.f-years to 32.1 for those with an index of 800 m.p.c.f-years. This important evidence of a dose-response relationship between the degree of dust exposure and the biological effects has considerable implications, both when considering methods of achieving safe working conditions and possible effects on non-occupationally exposed groups.

Table 6.2 Equivalent average death rate for lung cancer per 1000 men (Canadian chrysotile miners) (Macdonald, J. C., 1972)

Dust index (m.p. c.f.—years)	Average death rate
10	10.3
10—	13.1
100—	13.4
200—	15.5
400—	21.4
800—	32.1

Several attempts have been made to measure the size of the risk to the workers of developing lung cancer. It varies according to the degree and type of exposure, and to the smoking habits. In a study of shipyard workers in Northern Ireland about 25% of the deaths were from lung cancer; in the New York study of insulation workers 40%; but in the Canadian chrysotile miners and millers the figure was as low as 4%.

Cancers of the pleura and peritoneum—the mesotheliomata

Up to twenty years ago mesothelial tumours were regarded as very rare; indeed some schools of pathology denied that this was a definite pathological entity. The tumour which arises from certain cells of the pleura and peritoneum produces a complex and variable picture on microscopic examination; identification may be difficult, and confusion with tumours of primary origin in other organs is not uncommon. The association between exposure to crocidolite asbestos dust and these tumours was first noted in South Africa by Wagner and his colleagues, who published their observations in 1960.

A group of 33 patients with unusual features of chest disease, at first thought to be due to tuberculosis, was investigated. It was found that 32 of these patients had connections with the crocidolite mines in Cape Province of South Africa. There were both men and women, a large proportion had worked at the mines and others had played as children on the mine dumps or accompanied their mothers when they extracted fibre from ore in a process known as *cobbing*, where small asbestos rocks are broken up by a hammer. Others had been occupied in transport of crocidolite fibre by wagon from the mine to the railhead. The diagnosis in all cases was confirmed by autopsy and microscopic examination of the tumours, which in this group all affected the pleura.

A constant feature in this series was the long interval (average 20 years) between first exposure and development of the cancer. This period, known as the *lapse period*, is a common feature not only in asbestos-related cancer but in other cancers caused by industrial exposures to certain chemicals. The length of the lapse period adds considerably to the complexity of the problem, for by the time the individual becomes ill, he may have retired altogether, or have left the relevant occupation and have been employed in a variety of different jobs.

Table 6.3 Types of asbestos exposure of 76 patients dying of mesothelial tumours (data from Newhouse, M. L. and Thompson, H. (1965), *Brit. J. Industr. Med.*, **22**, 261)

Employed at asbestos factories	23	(30.3%)
Insulators and laggers	8	(10.5%)
Relatives worked with asbestos	9	(11.8%)
No occupational or domestic exposure but live within $\frac{1}{2}$ mile of asbestos factory	11	(14.5%)
No known contact with asbestos	25	(32.9%)

If there is a suspicion that a substance encountered at work is responsible for the development of a particular cancer, it is necessary to take a very detailed work history starting from the first jobs after leaving school; it is wise also to inquire about places of residence if an environmental influence is suspected. This technique was adopted in an investigation by Newhouse and Thompson of a group of 83 patients of the London Hospital in Whitechapel, in whom diagnosis of mesothelioma either of pleura or peritoneum had been confirmed by two pathologists experienced in this field. These patients had died between 1917 and 1964, the majority after 1955. In all but seven, a surviving relative was found who could give detailed information about the deceased (Table 6.3). There were 34 men and 42 women included in the series. 24 of them had worked in asbestos factories, 19 in one large factory in London's East End; one of these had delivered goods to the factory twice a week over a period of 11 years. All the other workers had been in departments of the factory which were known to be very dusty. This factory employed women workers in the spinning and carding sections where asbestos textiles were made. Eight of the workers were laggers. There was another group of 9, which included seven women whose husbands had worked with asbestos in the factories or unloaded it at the docks, and 2 men whose elder sisters had worked in the asbestos factory when they were boys. Asbestos dust had been brought into the home on the work clothes and hair of these people. There were a further eleven of the group who had never themselves worked with asbestos or had a relative with an asbestos job, but who had lived for some years, usually as children, within half a mile of the asbestos factory. This factory, which opened in 1913, was a heavy user of crocidolite asbestos. The duration of actual exposure in these patients varied between 2 months and 50 years. The lapse period varied between 16 and 55 years, with a mean of 37.5 years.

Both in the South African investigation and this later study there is definite evidence that non-occupational exposure, particularly to crocidolite, can be related to the development of mesothelial tumours; later studies in the United States and in Hamburg have confirmed these findings. Although development of the tumours may occur after short industrial exposure (there are not infrequent reports of asbestos exposure lasting only a few months), careful analysis of length and severity of exposure in relation to the numbers of tumours developing in persons with different categories of exposure has, as in the case of lung cancer, revealed evidence of a dose-response relationship. Those with long and heavy exposure are at far greater risk than those with short exposures in less dusty jobs. The mesothelioma rate/100,000 years at risk according to the type and length of exposure is shown in figure 6.3.

Because of the great length of the lapse period it is quite unsafe to assume that industrial conditions prevailing now or in the recent past can be equated with those prevailing thirty or forty years ago and that a neighbourhood risk still occurs in vicinities where asbestos is used, whether in factory or shipyard.

There is considerable evidence which links the development of

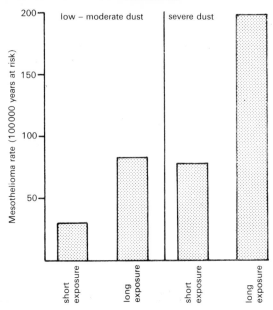

Figure 6.3 Mesothelioma rate per 100 000 years at risk by type and length of exposure

mesothelial tumour to exposure, particularly to the blue asbestos crocidolite. Wagner's original observations were made in the vicinity of the crocidolite asbestos mines of Cape Province of South Africa; since that time more than 300 miners from that area had died of the tumour. Among the 11 000 records examined in the chrysotile mining area in Quebec, only 7 deaths from mesothelioma were recorded. In Finland there were no deaths from this cause among 1000 anthophyllite miners. The contrast with the incidence of these tumours in crocidolite mining areas is striking.

In factories, exposure to more than one type of fibre is almost inevitable. Different products are made at different times, and economic considerations and availability of fibre from different areas may also influence the type of material which is used at any particular time. It is often difficult to determine the type of fibre to which an individual has been exposed in the past but, where a definite history is available, crocidolite fibre is far more frequently implicated than other types. In the United States, however, Selikoff has found a considerable number of deaths from mesothelial tumours in a factory population, chiefly employed during the second world war, where only amosite fibre imported from South Africa was used.

Experimental evidence from mesothelial tumours in rats, induced by intrapleural inoculation of measured doses of individual types of asbestos fibres, supports the theory that crocidolite has the greatest carcinogenic potential, followed by amosite, chrysotile and anthophyllite, in that order.

No co-carcinogenic factors have been implicated for this tumour in contrast to asbestos-related lung cancer. Its incidence is not related to smoking, nor is asbestosis invariably present. Immunological factors may be of importance, and methods of treatment based on stimulating immune responses are now being investigated with some improvement in survival time, though as yet no successful outcome has been achieved with any of the therapeutic regimes used in the treatment of other cancers.

Peritoneal mesotheliomata pose some difficult problems. The mode of access of the asbestos fibre to the peritoneum is not entirely clear, but asbestos fibre after inhalation may be coughed up in the sputum, swallowed, and then migrate through the bowel wall to the lining membrane (the peritoneum) or it may be conveyed by the blood vessels, or lymphatic channels from the chest cavity to the abdominal cavity. Another unexplained problem is why in some series, such as the insulators in New York and the London East End factory workers, peritoneal tumours are as commonly found as pleural tumours, whereas in other areas they are unknown. For example, in Scotland, 90 pleural mesotheliomata were identified, but no peritoneal tumours. There may be difficulties in accurate diagnosis due to confusion with other abdominal cancers, but this would not entirely account for the discrepancy in numbers, particularly in Scotland where a research group which was well aware of the problem was responsible for the investigation. The difference in incidence of tumours of the two sites may be due to differences in the amount of exposure, or differences in the type of fibre encountered by the workers. There is a field here for further research.

The size of the risk

It is difficult to assess the total mortality due to mesothelial tumours in asbestos workers, the available figures coming only from detailed studies of certain groups, where there is accurate knowledge of the size of the group, and the cause of death of each member has been verified. In many countries, however, the only information available comes from cancer registries where registration may be incomplete, or from the records of chest clinics and hospitals, where again information may be incomplete and inaccurate. The best available estimates come from Selikoff's work in New York, and from Newhouse's studies of the East End of London factory. Selikoff states that among his insulation workers 5% of the total mortality is due to deaths from mesothelial tumours. Newhouse calculates that the present mortality in the London factory group due to mesothelioma is approximately 7%, but by the end of the century it will probably rise to

11%. It will be remembered that the interval between first exposure and development of the disease is very long, but by the year 2000 more workers will have endured the lapse period and a higher proportion of deaths due to this cause can be expected.

Many detailed studies of the problem have been made in the United Kingdom, particularly by the Employment Medical Advisory Service, the medical branch of the Health and Safety Executive. In the whole country

Figure 6.4 Mesotheliomas associated with asbestos exposure up to 1973

over the years nearly 800 deaths from this tumour are known to have occurred (figure 6.4). The concentration of cases around ports, dockyards and industrial centres will be noted.

With the assistance of the Registrar General all death certificates mentioning mesothelioma are collected by the Employment Medical Advisory Service, thus maintaining a mesothelioma register. In 1969 there were 126 deaths due to this cause, rising to 190 in 1973. In about two thirds of the cases there was a history of exposure to asbestos dust.

Other asbestos-related tumours

It is always difficult to demonstrate that cancers which occur commonly in the general population are a particular risk of any one occupational group. Cancer of the lung is so markedly more common among asbestos workers than among the general population that statistical comparison of figures leaves no room for doubt and, similarly, mesothelial tumours exceedingly rarely occur in adults, except among those who have been exposed to asbestos. However, when gastro-intestinal tumours are considered, the figures are not so striking and there is also the possibility that any available figures have been distorted by inaccurate diagnosis; in at least two mortality studies where, whenever possible, the registered cause of death is verified by further investigations, asbestos workers are found to die of cancers of the stomach and bowel between two and three times more frequently than would have been expected from studies of the general population.

Cancer of the larynx is another tumour which is suspected of being casually related to asbestos exposure. At least two studies have shown that this cancer occurs more frequently than would be expected in men whose work brings them into contact with asbestos dust. It is already known that smoking (and probably alcohol consumption) is important in the causation

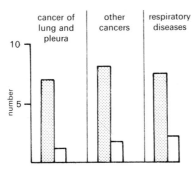

Figure 6.5 Number of deaths from various causes among severely exposed asbestos workers compared to a matched unexposed population.

M

of these cancers, and further studies are being undertaken to determine the relative importance of these three influences. Finally, cancer of the ovary has also been implicated. There is no firm evidence here, and there is a very strong likelihood that advanced cases of cancer of the ovary could be confused with peritoneal mesothelial tumours.

A graphic representation of the results of a long-term study of London asbestos factory workers is shown in figure 6.5. Approximately 150 of the group were first employed between 1933 and 1950. The dark columns show the number of deaths from cancers and chronic lung disease that have already occurred. The light columns show the deaths that would be expected to occur in a matched group of men who had no special occupational risk. These latter figures are calculated from data supplied by the Registrar General. The diagram summarizes very dramatically the excess risk which has been run by asbestos workers.

Mechanism of asbestos cancer production

It is generally accepted that the asbestos fibre is a powerful carcinogen, and the exact mechanism of cancer production, and the relative carcinogenicity of the different types of fibre has been the subject of active experimentation in laboratories over the past 10 years. These laboratory experiments have of necessity been performed on animals, usually rats or hamsters.

The type of experiment which has yielded the most informative results is the injection of various preparations of asbestos fibre into the pleural cavity of the animal. The number and type of tumours developing is then observed, usually over a period of two years. As with all animal experiments, there is difficulty in extrapolating the results to the human situation; nevertheless, important information has been obtained.

The first theory to be explored was that the asbestos fibre was contaminated by natural oils and waxes containing polycyclic hydrocarbons, known to be powerful human carcinogens. This theory was abandoned as experiments showed equal numbers of tumours arising in rats inoculated with crocidolite fibres, whether untreated or oil-extracted. The possibility that the carcinogenic activity was due to trace metals such as chromium was also explored, but no correlation was found between the number of tumours produced and the chromium content of the asbestos specimens. Experiments were also made with finely particulate metallic nickel, stainless steel and non-crystalline silicon dioxide, all of which could contaminate the fibre during the milling process, but none of these materials was found to be sufficiently carcinogenic to account for the asbestos effect. At present it is considered that the structural features of the asbestos fibres are the critical factor.

The external agents which contribute to the cause of cancer fall into one of three major groups: ionizing radiation, chemicals and viruses. Ionizing radiations need not be considered in this context; experimental evidence excludes the known chemicals associated with asbestos but particles that are within the dimensional range of viruses are abundant in all forms of asbestos; and it is conceivable that submicroscopic particles could penetrate the cell and act in a fashion similar to that of viruses. The actual change that occurs within the cell is still not yet understood.

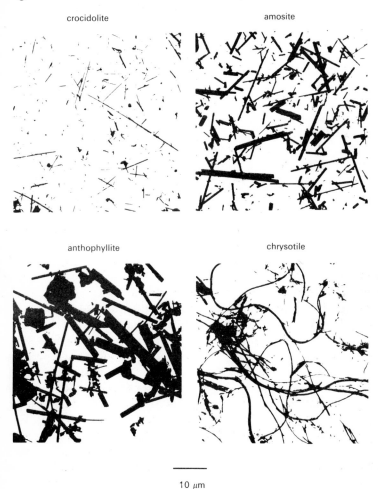

crocidolite

amosite

anthophyllite

chrysotile

10 µm

Figure 6.6 Electron-microscope photographs of four types of asbestos.

Study of the structure of the different types of fibre has helped to elucidate some of the mechanisms of their action in the lung. Figure 6.6

shows electron-microscopic photographs of the four types. Crocidolite has the shortest and thinnest fibres. Chrysotile is distinguished by the long curly fibre; amosite and anthophyllite both have thicker and straighter fibres. These characteristics influence both the degree of the fibres' penetration into the lung and their retention in lung tissue. Both animal experiments and studies in lung casts and narrow tubes suggest that the thinner shorter fibres of crocidolite penetrate deeper into the finest structures of the lung, and from there may migrate to the pleura. The amphiboles (crocidolite, amosite and anthophyllite) are retained longer in lung tissue than chrysotile. These important observations, which have helped to explain some of the apparent paradoxes in the varying incidence of asbestos-related tumours, have been carried out by a small group of scientists in Cardiff supported by the Medical Reasearch Council.

The risk to the community
Environmental pollution
First the evidence that any community, apart from occupational groups, has been affected by asbestos dust must be examined and, secondly, the possible effects on health should be assessed. The evidence began to appear in the mid-1960s. A pathologist in Cape Town decided to look for asbestos bodies (p. 140) in post-mortem examinations. A parallel investigation was carried out in Miami. In both investigations, asbestos bodies were found in about 30% of the male and 20% of the female lungs. These were adults, chiefly town dwellers; their occupations were not known, but there was no obvious connection with asbestos.

During the next few years similar investigations were made in centres in the United Kingdom, the United States and in Europe, and methods were devised of making accurate counts of the number of asbestos bodies seen under the microscope. In some investigations the previous occupations were determined. In general it was found that asbestos bodies were present in the lungs of up to 40% of urban dwellers, higher counts being found in men than in women; in people living in rural areas they were less frequently observed. Those with occupational exposures had the highest concentration of asbestos bodies in their lungs.

Recent electron-microscopy studies have revealed that a very high proportion of adult town dwellers have uncoated asbestos fibres in the lungs, but they are rarely found in children. The authors of a detailed study from the London Hospital comment that the actual number of asbestos bodies found is scanty when compared to the numbers usually seen in patients who had had asbestosis. There appeared to be a positive association with cancer of the stomach and cancer of the breast, but with no other cancers, including cancer of the lung. This series of investigations is a positive confirmation of asbestos exposure in large sections of the community. The source of exposure is often not known, but it may be from industrial sources, from building or demolition sites, or from brake shoes of motor-cars.

Further evidence of environmental contamination comes from radiological surveys of communities living in the vicinity of asbestos mines. The earliest of these surveys was made in Finland, in the neighbourhood of an open-cast anthophyllite mine. The chest X-rays of several hundred people showed evidence of calcified pleural plaques characteristic of asbestos exposure. There was an unexpected finding in Bulgaria when, in the course of an anti-tuberculosis campaign, chest X-rays showed typical asbestos plaques. It was found that the natural rock in the area contained a considerable proportion of anthophyllite asbestos.

Asbestos bodies and fibres in the lung, and plaques in the pleura are evidence of asbestos exposure, not of asbestos-related disease. However, there is evidence of mesothelial tumours causing death in people living in the vicinity of mines, asbestos factories and dockyards. The experience near the East End Factory in London has already been described. "Neighbourhood" tumours have also been identified in towns with big ship-building yards, such as Belfast and Hamburg, and in several American cities with asbestos industries. It must be remembered that the conditions which caused the deaths of people within the last 10 years were those prevalent twenty or more years ago. Exposure may have been considerable. One patient suffering from mesothelioma recalled playing as a small boy with handfuls of asbestos on waste ground near a factory some forty years before he developed his tumour.

Recent efforts have been made to measure the amount of asbestos in air in neighbourhoods where asbestos pollution is suspected. In New York, measurements were taken a short distance from a new building where concrete pillars were being sprayed with asbestos as a fire-proofing

Table 6.4 Chrysotile content of New York City air in vicinity of spray fire-proofing with asbestos-containing materials

Site	Asbestos level in ng/m^3
1. Downwind from source	45–180
2. 45 from source	15–30
3. Upwind from source	20
4. Downwind from source	45
5. Upwind from source	20
Industrial T.L.V.	120 000

measure. The measurements that were obtained are shown in Table 6.4. The highest recorded level was 180 ng/m^3; the level permitted in the air of British factories is 120 000 ng/m^3. (1 ng = 10^{-9}g.)

Asbestos in food, water and other beverages

Before considering the implications of such low-level exposure, mention should be made of the information that has accumulated about vehicles for ingestion of asbestos. Interest in this subject is high in view of the occurrence of abdominal tumours, i.e. mesothelial tumours of the peritoneum, and of the increased incidence of gastro-intestinal tumours in occupationally exposed groups. The first relevant scientific paper came from two workers at St. Bartholomew's Hospital in 1965. They were aware that asbestos pads were frequently used for filtration in the drinks industry. Using the electron microscope they identified asbestos fibres in water after it had passed through asbestos filters. They then examined samples of bottled and canned beer; fibres were clearly seen to be present and identified as chrysotile. It was estimated that a pint of the beer contained about 5000 fibres. This work was continued in Canada, where fibres were found in a variety of soft drinks, beers and sherries. Tap water in a number of Canadian towns was also examined; fibres were present in all, but in lower concentrations than in the English beer. In the tap water of Thetford, one of the asbestos mining towns, the number of fibres was specially high. The fibres probably came from asbestos ore in the areas from which the water supply was drawn.

There has been considerable concern in the United States recently, as the water supply of the town of Duluth is contaminated by effluvia from a large steel works which contains asbestiform fibres. There has also been anxiety that asbestos fibres from asbestos cement pipes might contaminate the water supply. This type of pipe has been in use for more than 40 years. Chrysotile asbestos has also been found in a variety of widely used drugs that are given by injection. Another investigation identified asbestos in rice prepared by a special method in the United States and exported to Japan.

The finished product

Anxiety is often expressed by members of the general public that manufactured goods used in the home, such as asbestos pads on ironing boards or mats used for cooking might be hazardous. In general the anxiety is unfounded since finished articles are usually sealed with a resin, and asbestos fibre cannot be liberated into the air. Nor do asbestos blankets (often kept in schools for fire extinguishing) liberate asbestos fibre when shaken. However these blankets (both for their fire-prevention function and as precaution against liberating asbestos fibres) should be discarded if frayed or damaged. The research group at Mount Sinai in New York identified an overcoat on sale, which stated on the label that 30% asbestos fibre was woven into its fabric. When the coat was vigorously brushed, a considerable number of asbestos fibres was liberated. Asbestos boarding is often used by do-it-yourself enthusiasts for building garages or garden sheds, or in other projects in the home. Sawing the board can liberate dust, and suitable precautions should be taken—such as working in the open air. Knowledge of a possible hazard leads to commonsense methods of avoiding it, and materials containing asbestos should be suitably labelled.

Assessment of the environmental risk

As has been shown, there is no lack of evidence that asbestos is present in the air breathed by many large communities, and without doubt asbestos fibres can be unknowingly ingested. Asbestos exposure is a serious risk to health, as has been shown by the study of industrially exposed groups. Is the population then at risk?

The chief argument of those most acutely concerned is that, although the concentration of asbestos in air in towns and cities is many orders lower then the concentrations found in factories, even in those with good control of dust exposure, the exposure to the community is throughout the 24 hours of the day and lasts from birth to death, whereas allowable concentrations in industry are based on an 8-hour day, 5 days a week, for a normal working life of up to 50 years. Secondly, a cause for anxiety is that asbestos fibre when in the tissues of the lung or other organ may be present and identifiable for an indefinite time. The exact mechanism which initiates a cancer is not known: one fibre in the lung may be sufficient.

The chief argument against this point of view is based on the concept of the dose response relationship. As has been shown, the risk of developing both asbestosis and the two tumours which are most definitely linked to asbestos exposure (lung cancer and mesothelioma) is certainly related to the intensity of exposure. People in factories, even many years ago when conditions were probably poor, who had short exposures in jobs with little contact with asbestos were about one sixth as likely to develop a mesothelioma as those with heavy and long exposure (figure 6.3).

Can there be a risk to those with less than one 10 000th of the industrial risk? There is further circumstantial evidence that cancer rates in the general population have not been influenced by the growing use of asbestos. Current research has only identified clusters of mesothelial tumours in areas where asbestos is used in large quantities—in mining areas or in ports, dockyards or centres of manufacturing. Calculations of the mesothelioma rate in all states of the United States and Canada showed an increase only where there are asbestos mines or an asbestos industry. No evidence has been produced by the very many research groups in the United Kingdom, Europe or in the United States that the lung-cancer rate is raised where there is possible asbestos pollution from industrial sources.

An authoritative report came from the International Agency for Research on Cancer, a branch of the World Health Organisation, in 1972. The authors did not believe that there was a risk of either lung cancer or mesothelial tumours to the general population from asbestos in the air or contaminated water, and they accepted that mesothelial tumours could occur, but rather infrequently, without asbestos exposure. They did not find any evidence that fibrosis or lung cancer was caused by environment contamination.

Future policy and research
There is no doubt that the modern world needs asbestos. It seems unlikely that any material or variety of materials will be developed that can substitute for its great diversity of desirable properties. On the other hand, modern industry and technology must not exact unacceptable health risks from those concerned in the primary production of the materials required.

There are clear indications that the different types of fibre exert varying influences on human cells, but little is known of the different behaviour of the fibres when in the body tissues. More basic research might reveal the properties which lead either to fibrosis or cancer production, and identify how asbestos might be modified to avoid these properties. There is also substantial evidence that at certain levels of exposure the risk of ill effects is so low that it becomes acceptable—but opinions vary about the level of exposure safe for the normal working man and, by extension, the community at large. Great Britain in 1968 adopted a standard of 2 fibres/ml in the working environment as one that would cause only a one-in-a-hundred risk of pulmonary fibrosis in a normal working life. At that time there was insufficient evidence to suggest a standard that might reduce the risk of cancer to the same level. Recently in the United States recommendations have been made to reduce the standard there to one quarter of the British standard.

More technological research must be done to find materials with a similar performance to asbestos; some progress has been made. Glass wools, fibre glass and other materials are being used in insulation, but at very high temperatures no substitute has been found for asbestos lagging. Asbestos substitutes generally have a poorer performance, less durability and are more expensive. A note of warning against using unproven materials was heard when a Californian scientist reported the occurrence of mesothelial tumours when glass fibre reduced to the dimensions of asbestos fibrils was inoculated into the pleural cavities of rats.

Other methods of control rest on improved technology in the factory. The dangerous process must be isolated and enclosed, and as far as possible automated so that few workers are in contact. Effective methods of dust suppression and dust extraction must be employed to keep atmospheric dust at an acceptable level; where escape of asbestos fibres into the air cannot be prevented, the worker must be given personal protection, either well-fitting dust mask or even, in certain situations, a clean air supply through ventilated hoods or helmets. Situations in which the control of dust prevents very difficult technical problems are not very common, but they may arise when old lagging has to be removed. The precautions necessary are shown in figure 6.7.

Figure 6.7 Stripping old insulation. The operative is fully protected. The insulation is thoroughly wetted and polythene sheeting is used to avoid deposition of asbestos in inaccessible places

Factory hygiene is the responsibility of the owner, but standards are controlled by statutory regulations; in the United Kingdom, all factories and other places where asbestos is used are open to inspection by the Factory Inspectorate. There is also medical supervision of all workers. Every worker must be examined and have an X-ray of the chest every second year. Careful records are kept of the type of job and the level of dust in the air in which he has been working. Medical examinations, X-rays and analyses of the data from the examinations of more than 25 000 workers is undertaken by the Employment Medical Advisory Service. The scheme serves a double purpose—first, the early detection of adverse changes in any individual, so that advice can be given and treatment instituted and, secondly, evaluation of the efficacy of present standards and regulations. The first asbestos regulations in Britain were made in 1931. At that time there was only a small industry, chiefly producing fire-proof textiles; the only recognized medical consequence of asbestos exposure was lung fibrosis. As the industry expanded and diversified, further medical risks were identified, but the regulations and standards were not revised till 1969. With the current awareness of the dangers by industry, government bodies, trade unions and the general public, stricter monitoring will be undertaken. The disablement and loss of life of asbestos workers during the past decade is due to conditions in the earlier part of the century. The lessons of this tragic experience must be learnt by all countries, and particularly by developing countries with new asbestos industries. Other industries, and in particular the chemical industry, which has had its own bitter experience with a chemical carcinogen now banned from manufacture and import, should also develop adequate methods for short-and long-term monitoring of the health of the worker.

FURTHER READING

Biological Effects of Asbestos. Annals of New York Academy of Sciences, 1965, Vol. 132.
Biological Effects of Asbestos, IARC Scientific Publications No. 8, International Agency for Research on Cancer, Lyon, 1973.
Parkes, W. Raymond (1974), *Occupational Lung Disorders,* Butterworth.

Index

Acetarsol 100
acid rain 21
activation analysis 95
aerosol sprays, effect on ozone equilibrium 14
aflatoxins 121-136
  and hepatitis 129, 130
  carcinogenic properties 127, 130-135
  chemical composition 125
  conditions for growth 124
  effect on liver 128
  in milk 125
  in nature 123
  interaction with DNA 127
  LD_{50} values in animals 128
  metabolism 126
  stability under heat 126
  toxicity in animals 127
    in man 128
Agricola 115
agricultural productivity 12
air pollution by oxides of nitrogen 19
Albertus Magnus 99
alkoxyalkylmercurials 48
alkylmercurials, toxicology 48
aminolaevulinic acid (ALA) 86
ammonia 16
  in soils 17
anaemia 80
anaerobic bacteria 24
antibiotics 121
anti-knock agents 29
arsenic 31, 93–120
  affinity for hair 97
  alleged aphrodisiac power 98
  and cancer 115, 116
  as growth stimulator in pigs and poultry 101
  as instrument of homicide 109
  cycle 33
  daily intake 97
  health hazards 34
  in beer 96, 106
  in body tissue 98
  in cigarettes 118
  in coal 96

  in defoliants 96
  in food 97
  in hair 110
  in pesticides 96
  in rocks 95
  in the environment 95
  in treatment of syphilis 100
  in wine 97
  industrial production 32
  legislation restricting sale 99
  medical uses 99
  not an essential element in man 98
  poisoning 99, 102, 105
    accidental 111
    in animals 119
    in industry 107
      decline since 1900 108
      in vineyard workers 104, 109, 118
  pollution of water supply 109
  toxicity dependent on valency 96
  trichloride 107
arsenical cancer 117
  insecticides 101
arsine 101, 112
  poisoning 104, 112, 113
    at sea 114
    decline since 1900 113
    in baloonists 112
    in industry 111
arsphenamine 94, 100, 119
arylmercurials 47
asbestos 137–158
asbestos and cancer 141, 156
  and factory hygiene 157
  biological effects 139
  bodies 139
  corns 139
  dose response relationship 155
  environmental risk 155
  in air 153
  in beer 154
  miners, lung cancer mortality 143
  plaques 141
  safe levels of exposure 156

workers, increased mortality from various
  diseases 149
    relation between smoking habits and
      mortality 142
    world output 139
asbestosis 140
Aspergillus flavus 123, 127
atmosphere, residence times of various
      constituents 3
    of carbon dioxide 10
    of nitrogen dioxide 18
    of nitrous oxide 18
    of particles 3
    of sulphur dioxide 20
  scavenging by rainfall 3
atomic absorption spectrophotometry 44
atoxyl 94, 100, 119
Avicenna 99

bacteria, sulphur production by 24
Barth 2
basophilic stippling 80
Belmonts Balsam 99
biological cycling 6
biosphere, inputs and outputs 7
black spot in roses 21
blood lead in urban and rural environments 89
blue asbestos 137
Brazil nuts as source of aflatoxin 124

cadmium 31
  cycle 33
  health hazards 34
  industrial production 32
cancer production in asbestos workers,
    mechanism of 150
carbon-14 13
carbon cycle 8, 9, 12
carbon dioxide and plant growth 10
    increasing in the atmosphere 9, 10
    influence on climate 11
    residence time in the atmosphere 10
carbon monoxide 12
Celsus 98
Charles the Bad 99
Chaucer 99
chelating agents 92
Cheyne, George 38
chromium 31
  cycle 33
  industrial production 32
chrysotile 137
  uses 138
cider, contamination by lead 68
cinnabar 39

cirrhosis and liver cancer 133
climate, influence of carbon dioxide 11
coal and oil resources 11
    as source of arsenic 31
cobbing 144
Collis 105
copper cycle 28
  in lake sediments 30
  in soil 30
  industrial production 28
corrosive sublimate 46
crocidolite 137
  uses 138
cycling of elements in the environment 6

DA 94
DC 94
demethylation of mercury 51
depletion of natural resources 8
Devonshire colic 68
dimethyl mercury 34
dinitrogen 16
Dioscorides 67, 98
dioxygen 13
DNA interaction with aflatoxins 127
Doig, A.T. 113
dose-response curve 45
    relationship for asbestos 155
dynamic equilibrium 1

Ehrlich 100
electron microscopy in study of asbestos
    hazards 140, 151
encephalopathy 82
environmental pollution by asbestos 152
estuaries, residence times 4
estuarine muds 5
eutrophication 17, 26, 27

ferrochelatase 80
ferro-silicon 114
fertilizers 17, 25
fixation of nitrogen 16
fluoride, residence time in ocean 27
Fowler's solution 100, 104, 116
French Green 101
fuel oils, sulphur in 20, 24
    reserves 11
fungi as sources of antibiotics 121

Galen 98
gas chromatography 44
Geber 99
geological cycling 6
globin 80

glomerulus 82
grain and aflatoxin poisoning 130
Great Lakes 27
Gutzeit's test 95

Hadrian's Wall and lead mines 67
haem synthesis 80, 86
haemoglobin 80
haemolysis 80
half-life 2
heavy metal cycles 28
Hippocrates 67
Hutchinson, Jonathan 108, 116
hydrogen fluoride, health hazard 27
 sulphide 20, 24

inert gases 2
itai itai disease 34

Joachimsthal 117

lake sediments 24
Lake Washington 4
lead 64–92
 absorption from gut 70
 aerosols 30
 alkyls as petrol additives 29
 and ALA 86
 and hypertension 90
 and neural retardation 89
 and opium 76
 and protein synthesis 87
 arsenate 97, 101
 biochemical effects 83
 colic 67
 concentrations in blood and urine 89
 cycle 28, 69
 deposition in bone 70
 distribution in teeth 70
 effect on enzyme activity 85
  on foetus 90
  on kidney 82
  on nervous system 82, 87
 excretion 70
 global contamination 30
 health hazard 69, 75
 in air 74
 in blood, correlation with lead in water 73
 in brain 71
 in food 73
 in lake sediments 30
 in moonshine whisky 78
 in soil and plants 30, 74
 in water 73, 78
 industrial exposure 77

industrial production 29
 no essential metabolic function 84
 pathological effects 80
 piping, health hazards 67
 poisoning 69, 78
  acute and chronic 76
  decline since 1900 70
  in animals 75
  prevention 91
  sub-clinical 88
  treatment 91
 principal uses 68
 recommended limits of intake 74
 sources of exposure 73
 technology in Rome and Greece 67
 uptake and distribution in body 72
  from water pipes 74
leaded petrol 79
Legge, Thomas 116
Lenihan, John 119
lewisite 94
lichens, damage by sulphur dioxide 22
ligands 85
Liquor Arsenicalis 100
liver cancer, geographical distribution 131–5
 in man 131
 in trout 121, 123
 effect of aflatoxins 128, 129, 130
Los Angeles 18
Luce, Claire 105
lung cancer 117

Maimonides 99
marine sediments, residence time 5
Marsh's test 95
mass spectrometry 44
Maybrick, Mrs 109
mercurial pesticides 54
mercury 31, 38–63
 acceptable daily intake 54
 air quality standards 60
 analytical methods 44
 background level 56
 commercial production 42
 content of atmosphere 43
 cycle 33, 41, 43
 degassing of earth's surface 42
 development of tolerance 52
 emission standards 60
 environmental standards 55
 genetic damage 52
 health hazards 45, 46
 in air 58
 in coal 35
 in earth's crust 39
 in food chains 53

in foods 59, 61
in treatment of syphilis 46
in soil 35
in water 57
industrial production 33, 35
intake from food 53
nature's protective mechanisms 51
poisoning 46
release from coal burning 42
    from earth's crust 42
    to environment 42
Merriman, Seton 105
Mersey estuary 5
mesothelial tumours in asbestos workers
    143–149
metals, residence times in water 4
methyl mercury 34, 59
in fish 48
poisoning 50
methylation of mercury 50, 51
Minamata 48, 50, 51, 52
moonshine whisky, lead in 69
mutation rate 13
mycotoxicosis 130

Napoleon, arsenic poisoning suspected 107
natural cycles, effects of human activity 8, 10,
    36
resources depletion 8
neutron activation analysis 44
Nicander 67
nickel 31
carbonyl 117
cycle 33, 35
health hazard 36
in soil 35
industrial production 33
Niigata 48
nitrates
conversion to nitrogen 17
fertilizers 17
health hazards 17
in soils 17
industrial production 17
nitric oxide 18
production in atmosphere 18
    in sea 18
    in soil 18
    in stratosphere 18
Nitrobacter 17
nitrogen cycle 15
dioxide 18
fixing bacteria 16
residence time in atmosphere 2
Nitrosomonas 17

nitrous oxide 18
residence time in the atmosphere 18

ocean, residence times in 5
Old Testament, references to lead 67
olive oil 67
Oliver, Thomas 98
organic arsenicals 100
organomercurials 35, 47
tolerance limits 54
orpiment 93, 94, 100
oxides of nitrogen 18
oxygen cycle 13
production by plants 13
residence time in atmosphere 2
ozone 14, 15
amount in atmosphere 14
effect on plants 15
in atmosphere, destruction by pollutants 14
residence time in atmosphere 14

Paracelsus 92, 115
Paris, Ayrton 115, 119
Green 101
Parkes process 66
peanut meal 122, 129
Penicillium rubrum 123
peripheral neuritis in arsenic poisoning 104
perturbations of elementary cycles 7
petrol additives 29
phenylmercurials, toxicity 47
phosphates, residence time in atmosphere 25
use in agriculture and industry 26
phosphine 24
phosphorus cycle 24, 25
photosynthesis 7
pica 78
Pliny 98
plumbism 69
productivity, agricultural 12
protection from arsenic hazards 119
protoporphyrin 80
pyrites 93

radioactive tracers 1
rainfall, scavenging atmosphere 3
rate constant 2
realgar 93, 94
red cells, life span 80
Reinsch 94
residence time 1–6
in atmosphere 2
    carbon dioxide 10
    particles 3
    sulphur dioxide 20

in estuaries 4
in ocean 5
　　copper 28
　　fluorides 27
　　lead 28
　　phosphates 25
　　zinc 28
in soil 6, 31
in water 4
Reye's syndrome 129
Rhases 99
rice and aflatoxin poisoning 130
rivers, speed of purification 4
rock phosphates 25
Rönnskär disease 108

salt domes 24
saturnism 69
Scheele 112
Schneeberg 117
semi-conductors, arsenic in 101
Shaw, G. B. 77
sheepdip 107, 117
skin cancer 115, 116
sleeping sickness 100
Smith, Hamilton 119
　　Madeleine 109
smog 18, 24
sodium arsenite 101
soil, pollution by metals 30
　　residence times 6
　　　various metals 31
Stoneleigh murder 109
sulphur cycle 19
　　dioxide as pollutant 21

concentration in air 23
damage to plants 23
residence time in atmosphere 20
seasonal variation 21
in fuel oils 20
production by bacteria 24
sulphydral groups 85
swordfish, mercury in 49
syphilis 46, 100

tests for arsenic 94
Thackrah, Charles 105
tree rings 30
trypanosomiasis 100
Tryparsamide 100
tuna, mercury in 49
turkey X disease 121, 122

Vitruvius 67
volcanic emission 20

water, pollution by arsenic 109
　　residence times in 4
white arsenic 94, 99
　　asbestos 137
Woodall, John 38

yellow rice disease 129

zinc cycle 28
　　in lake sediments 30
　　in soil 30
　　industrial production 29
Zürichsee 27